TEN PROPOSITIONS

REGARDING SPACEPOWER

BY

M.V. SMITH

A THESIS PRESENTED TO THE FACULTY OF

THE SCHOOL OF ADVANCED AIRPOWER STUDIES

FOR COMPLETION OF GRADUATE REQUIREMENTS

SCHOOL OF ADVANCED AIRPOWER STUDIES

AIR UNIVERSITY

MAXWELL AIR FORCE BASE, ALABAMA

June 2001

Disclaimer

The conclusions and opinions expressed in this document are those of the author. They do not reflect the official position of the US Government, Department of Defense, the United States Air Force, or Air University.

About the Author

Major M.V. Smith was born in North Conway, New Hampshire in 1964. He graduated with honors from Kennett High School in 1982 and Cum Laude from Saint Michael's College, Vermont, with a Bachelor of Arts degree in Psychology in 1986. He was a Distinguished Graduate from the Air Force Reserve Officer Training Corps and received a regular commission upon graduation. In 1987, Major Smith was a Distinguished Graduate of Undergraduate Missile Training and in 1994, he was a Distinguished Graduate from Undergraduate Space Training. He has served in various missile and space positions, including missile combat crew commander, instructor and evaluator, wing executive officer, missile maintenance officer, and space crew commander, trainer, and evaluator of several systems. In 1996 Major Smith was a Distinguished Graduate of the US Air Force Weapons School-Space Division where he later served as an instructor teaching space force employment of National and Department of Defense space systems into strategic, operational, and tactical war planning and execution. During Operation Allied Force, he served in the Combined Air Operations Center at Dal Molin Air Base, Vicenza, Italy, where he served in the air component Strategy Cell and on the Guidance, Apportionment, and Targeting team helping to shape the air campaign and the subsequent peacekeeping operation. Major Smith holds three graduate degrees including Master of Arts in Political Science from the University of South Dakota, Master of Military Operational Arts and Science form the Air Command and Staff College, and Master of Airpower Arts and Science from the School of Advanced Airpower Studies. In July 2001, Major Smith became a strategist assigned to the Air and Space Operations Directorate at Headquarters Air Force at the Pentagon.

Acknowledgments

This study was difficult for four reasons. First, there exists only a relatively small body of writing about spacepower—especially spacepower theory. Second, many works about spacepower are typically advocacy pieces wherein authors argue for a preferred organizational structure for stewarding American spacepower or future procurements, rather than critically discussing what spacepower brings to modern warfare. Third, there is some risk associated with writing about spacepower because doing so is at odds with the current Air Force policy that insists that air and space form an indivisible aerospace continuum. Finally, each proposition presented in this study becomes a separate thesis supporting the larger purpose of this study.

Special thanks go to my thesis advisor, Lt Col Peter Hays, and reader, Lt Col Forrest Morgan. The conversations we had not only helped me get my arms around this project, but also furthered my appreciation for spacepower's evolution. Col Phillip S. Meilinger, author of *Ten Propositions Regarding Airpower*, provided sage advice at the outset of this study. Considerable help also came from other SAAS faculty, especially Dr Karl Mueller, Dr David R. Mets, Dr Harold Winton, Col Dennis Drew, and Maj John Terino.

Interviews, lectures, and seminar sessions with several general officers helped me understand the organizational, political, and bureaucratic constraints imposed upon our national spacepower as the Air Force wrestles with its future and the concept of "aerospace." Among them were Gen "Chuck" Horner, Gen Ralph "Ed" Eberhart, Gen John Jumper, Gen Howell M. Estes III, Gen Joseph Ashy, and Gen Richard Myers. Lt Gen Glen "Wally" Moorhead and Lt Gen Lance Lord helped, as did Maj Gen Lance Smith, Maj Gen Howard Mitchell, and Maj Gen John Barry.

My brother space weapons officers also deserve considerable credit for helping me frame the propositions over the past three years. Lt Col "Mr Bill" Billman, and Maj "Scout" Kinnan remain my constant mentors and stand out as significant contributors. Lt Col Greg "Chappy" Chapman also provided valued comments from the Space Division at the USAF Weapons School. Lt Col Michael "Mik" Beno and Maj James Cashin from the Air Command and Staff College provided enthusiastic encouragement for this project and provided much needed reviews of early drafts. Over the past year, Maj Fred "Zelder" Gaudlip, a true spacepower advocate, and Maj Rob "Snort" Givens, a believer in the "vertical flank," helped me hammer-out many of the arguments presented here. Maj Scott Cook, from the USAF Doctrine Center, spent several hours with me engaged in very productive discussions. My wife, Alice, also deserves considerable credit. She endured many lonely hours while I worked and put up with me while I bounced ideas off her…over-and-over.

Most of the credit, however, goes to Nate—a little boy who is still too young to understand why Daddy closes his door. He paid the greatest price for this work. We are both looking forward to spending more time together.

Abstract

This study rides the coattails of Colonel Phillip S. Meilinger's book, *Ten Propositions Regarding Air Power*. As the United States ponders its future regarding space operations, the time has come to frame similar propositions regarding spacepower. Specifically, this study seeks to answer the question, "What is the nature of spacepower?" It also tests the aerospace integration school's hypothesis that spacepower is simply a continuation or extension of airpower. Two points come immediately to the forefront of this work. First, spacepower is different from airpower even though both share the vertical dimension of warfare. Second, space operations have matured to a point wherein valid and unique propositions regarding spacepower are identifiable. The ten propositions presented here do not represent a complete list. The method used to derive these propositions involved literary research that resulted in a long list. The list evolved over three years during numerous brainstorming sessions with several space experts— most of them space weapons officers with theater and often combat experience—until the list was carefully refined into the ten most salient propositions. There were many ways to present the Ten Propositions Regarding Spacepower, but the author deferred to Meilinger's approach of citing each proposition as a thesis statement with supporting material immediately following. The objective of this work is to stimulate discussions and help those who do not yet understand or appreciate the nature of spacepower in modern warfare.

Contents

Illustrations

Chapter 1

Introduction

Space, to a large extent, is an unknown to many throughout our country and to many leaders in our government who are being asked to make critical decisions that will chart the course of space for the United States-- both inside and outside the military.

—General Howell M. Estes, III

One should bear in mind that there is nothing more difficult to execute, nor more dubious of success than to introduce a new system of things: for he who introduces it has all those who profit from the old system as his enemies, and he has only lukewarm allies in all those who might profit from the new system.

—Machiavelli, "The Prince"

Either you are a separatist or a conformist. The separatists will often be killed by the party faithful; the conformist will kill the very organization they seek to defend.

—General Charles Horner
Former USCINCSPACE

The objective of this work is to assist those who do not yet understand or appreciate the nature of spacepower. Its aim is to help political and military leaders, practitioners of war, and interested citizens to better understand the nature of space in order to fully exploit its use as a source of national and military power. American airmen in particular will benefit from pondering the discussions contained herein as they wrestle with their evolving role in space operations. At the heart of this study lies the question, "What is the nature of spacepower?" The propositions presented in this study—along with supporting arguments—provide an answer to this question.

1

Significance of this Work

At the turn of the twenty-first century, American spacepower is on unsure theoretical and doctrinal footing. Despite more than forty years of spacefaring experience, there is still no great book about spacepower—no Clausewitz, Mahan, or Douhet.[1] Consequently, The United States Air Force (USAF) has yet to arrive at a definitive way to conceptualize space. It vacillates between the terms "aerospace" and "air and space" within its lexicon to describe the operating environments beyond the surface of the Earth. Indeed, this distinction gives rise to a heated and ongoing debate between members of two schools of thought within the Air Force.[2] It boils down to a political and organizational debate regarding whether or not space operations should remain in the Air Force or evolve into a separate service. The aerospace advocates view space as the manifest destiny of airmen, whereas the air and space advocates believe space is the manifest destiny of an independent space force.

This study will help train the judgment of military practitioners by pointing out the unique considerations of spacepower. At the same time, it will help readers to assess the strengths and weaknesses of the two competing schools and decide which side of the debate they favor. Readers are free to consider spacepower as a standalone concept, or as one of the three pillars of "aerospace power," which also includes airpower and information power.[3] Regardless of how the Department of Defense (DoD) organizes,

---

[1] Perhaps no great book about spacepower is possible, in the classic sense. Contemporary authors must meet higher academic standards than the great theorists. Their works would likely be doomed by modern editors for failing to cite their sources and resting their arguments on untested theory—often without using historical analogies as proofs.

[2] To witness the debate in progress, refer to *Aerospace Power Journal (previously Air Power Journal* prior to Winter 1999), particularly the following editions; Spring 1999, Summer 2000, and Spring 2001.

[3] Frederick L. Baier, *Fifty Questions Every Airman Can Answer*, (Air Force Doctrine Center, Maxwell Air Force Base, Ala. October 1999), 3.

trains, and equips to provide military spacepower for the nation, there are fundamental propositions regarding spacepower that persist in the face of bureaucratic, organizational, and political wrangling. This study identifies ten of these propositions and argues their case. These propositions may also serve as a foundation beyond the classical theories of statecraft and warfare to help other authors create spacepower theory, doctrine, and strategy.

This study unabashedly rides the coattails of Colonel Phillip S. Meilinger's book, *Ten Propositions Regarding Airpower*. While Meilinger's propositions are controversial, his book is nonetheless widely read. It stimulates healthy debates about airpower. Spacepower deserves the same attention, particularly as America pits its desire to preserve space as a peaceful sanctuary against its fears of foreign aggression, which may ultimately lead to the weaponization of space.

Definitions

Words mean things. This is especially true when framing propositions. Unfortunately, there is no fixed and time-honored definition for many of the terms used regarding spacepower, but in all fairness, most military terminology is also in a dynamic state of flux. Authors tend to use specific terminology differently. This places a burden on the reader who must exert some effort to fully grasp the meanings used by each author. Therefore, it is necessary to provide definitions to help guide the reader during the discussions that follow.

Proposition

Before defining what a proposition is, it is essential to explain what a proposition is *not*. First, a proposition is *not* a principle of war. The principles of war apply to all forms of military power, not just spacepower.[4] Second, a proposition is *not* a tenet. A tenet is an enduring belief about how to employ a particular form of power that rises to the level of institutionalized doctrine.[5] This is very similar to a proposition, except a proposition does not focus on employment, nor does it meet the criteria for doctrine. Third, a proposition is *not* a core competency. Core competencies are basic areas of expertise for practitioners within a form of power.[6]

For the purpose of this study, *a proposition is a statement suggested for acceptance.* This is consistent with the definition used by Meilinger.[7] The reader is free to accept or reject any proposition based on its own merits. It is possible that political or technical changes over time will supercede these propositions or their supporting arguments.

Power

The term "power" has many meanings. A precise definition depends on who is using it. For the purpose of arriving at a suitable definition of spacepower, *"power is the ability of a state or non-state actor to achieve its goals and objectives in the presence of other actors on the world stage."*[8] This definition does not require the actors to be states, nor does it require a precondition of competition between actors for power to exist in the global system.

---

[4] Air Force Doctrine Document (AFDD) 1, *Air Force Basic Doctrine*, 1 September 1997, 11-21.
[5] Ibid., 21-27.
[6] Ibid., 27-34.
[7] Col Phillip S. Meilinger, Naval War College, interviewed by the author, 20 July 2000.
[8] James L. Hyatt, III et al., "Space Power 2010," Research Report n. 95-05 (Maxwell AFB, Ala.: Air Command and Staff College, May 1995), 5.

Space

There is no approved definition of space, and a formal definition is probably not forthcoming. From the genesis of space flight, neither the United States nor the Soviet Union wished to define the term space for fear of constraining their activities in this new environment.[9] Instead, both nations elected to use a functional definition that *whatever is in orbit is in space.*[10] Nevertheless, many people develop their own definitions of space for the purpose of clarification. For example, Air Force officers entering today's equivalent of Undergraduate Space Training in the early 1990s read the following in their basic text:

> If trying to define where space begins for biological reasons, one might choose 9 miles above the Earth since above this point a pressure suit is required. If concerned with propulsion, 28 miles is important since this is the limit of air-breathing engines. For administrative purposes, one might find it important that US astronaut wings [are] earned above 50 miles. An aeronautical engineer might define space as starting at 62 miles above the Earth's surface since this is where aerodynamic controls become ineffective. Conventional and customary law defines the lower boundary of space as the lowest perigee of orbiting space vehicles, about 93 miles.[11]

For the purpose of defining spacepower in this study, using the conventional and customary definition is preferred; *space begins at the lowest perigee of an orbiting satellite, about 93 miles beyond the Earth's surface, and extending out to infinity.* This is essentially the functional definition, which all concede; if it is on orbit, it is in space. It includes objects that are not in orbit, but that achieve altitudes that may interfere with objects in orbit, such as ballistic missiles in transit. However, this definition is at odds

---

[9] Walter A. McDougall, *The Heavens and the Earth: A Political History of the Space Age* (New York: Basic Books, 1985), 180, 259.

[10] John F. Graham, *Space Exploration: From Talisman Of The Past To Gateway For The Future* (http://tycho.space.und.edu/projects/book/index.html) 11.

[11] Captain Carol Laymance, "Science of Space," in *Space and Missile Orientation Course* (Vandenberg AFB Ca.: 30th Operations Support Squadron, 1993), 1-3, cited in Hyatt et al., 93.

with the official USAF position, which claims that it is impossible to divide air from space because there is no distinguishable barrier between the two since the atmosphere trails off so gradually. Still, the Air Force concedes that if an object is on orbit, it is in space.

Spacepower

For the purpose of this study, *"spacepower is defined as the ability of a state or non-state actor to achieve its goals and objectives in the presence of other actors on the world stage through…exploitation of the space environment."*[12] This definition is remarkably similar to a definition for any other form of power, be it air, land, sea, or information. In the broadest sense, spacepower includes all activities performed by an actor—or exploited by an actor—in the space environment for civil, military, commercial, or other reasons.

Air

Air requires definition because many of the arguments used in this study refer to the air medium. *Air is defined as the area extending upward from the Earth's surface to an altitude where air-breathing engines can no longer operate, approximately 28 miles.* This definition is also at odds with the official USAF policy, which recognizes no upper limit to the air medium for the same reason it does not recognize a lower boundary for space. Ironically, the highest-flying aircraft in the Air Force's operational inventory, the U-2, only soars to an altitude approximately 16 miles above the surface.

Notice that the definition of air offered here is also a functional definition. When functional definitions of air and space are used, it becomes apparent that air and space do not meet. Airmen claim that air and space are a seamless continuum because it is impossible to identify a discrete altitude where air suddenly ends and space begins. While it is true that no discrete altitude divides air from space, it is really quite irrelevant. Between the ceiling of aviation and the floor of astronautics, there is a region nearly sixty-five miles wide that divides air and space. This is the *transverse region*, wherein neither aerodynamic flight nor orbital rotation is possible. Despite rhetoric to the contrary, the transverse region divides air operations from space operations, and removes the possibility of an "aerospace continuum."

Operations inside the transverse region are not practical because the energy expenditures required to maneuver there are too great. Vehicles can exploit neither Bernoulli's aerodynamic principles nor Kepler's astrodynamic principles to maneuver or conserve energy. Consequently, the transverse region lives up to its name as a boundary across which vehicles travel, but vehicles can do little else there. The great cost of space operations hinges on spacelift vehicles—large, fiendishly expensive rockets—that generate the huge amounts of energy required to lift payloads through the transverse region and accelerate them to orbital speed and altitude.

Aerospace

The term aerospace arrived on the scene in 1958 when General Thomas D. White first argued that air and space are indivisible and thus claimed space as the natural realm of the Air Force.[13] The Air Force has used the term on and off in various editions of doctrine and other publications. The other Services viewed this as a bureaucratic attempt

---

[12] Hyatt, 5.

[13] Thomas, D. White, Gen, USAF, "Air and Space are Indivisible," *Air Force*, March 1958, 40-41. For an in-depth discussion of the roots of aerospace, see Maj Stephen M. Rothstein, *Dead on Arrival? The*

by the Air Force to lay claim over a greater share of the future defense pie. As Hays and Mueller point out, "the other Services and the Office of the Secretary of Defense have never accepted the Air Force's definition of aerospace and have certainly not ceded all operations in this realm to the Air Force."[14] Not surprisingly, the term is missing from the Report of the Commission to Assess United States National Security Space Management and Organization United States, submitted to Congress on 11 January 2001, except in a reference to industry.[15] Nevertheless, the term was never more in vogue inside the Air Force than at the turn of the millennium, and never was it more controversial.

The term is evolving. Traditionally used as a noun, it is synonymous with "air and space," as if they are one in the same. In 1959, the Air Force defined aerospace as "an operationally indivisible medium consisting of the total expanse beyond the Earth's surface."[16] Even doctrinaires find this form of the word confusing because it ignores the obvious differences between air and space, often resulting in gross generalizations of the characteristics of one to the other.[17] Increasingly, aerospace is used by the Air Force as an adjective, as in, "of or pertaining to the total expanse beyond the Earth's surface."[18] Although this is a matter of semantics, the use of this subtle adjectival form of aerospace allows separate treatment of air and space under the umbrella concept of aerospace. It becomes a term much like maritime, which the Navy uses to refer to operations by ships

---

Development of the Aerospace Concept, 1944-58 (Maxwell AFB, Ala.: Air University Press, November 2000) 44-58.

[14] Lt Col Peter Hays and Dr Karl Mueller, "Boldly Going—Where?" *Aerospace Power Journal* 15 no. 1, (Spring, 2001), 36.

[15] This commission is commonly known as the "Space Commission," chaired by now Secretary of Defense Donald Rumsfeld. (Henceforth the Space Commission.)

[16] Air Force Pamphlet (AFPAM) 11-1-4, *Interim Aerospace Technology Reference*, 30 October 1959.

[17] Maj Shawn Rife, "Does Aerospace Mean 'Air and Space,'" n.p.; on-line, Internet 12 February 2001, available from http://sac/saic.com/space/docs/does.htm.

at sea or Marines ashore—a term that accommodates separate services and joint operations.[19]

Aerospace Power

It is best-defined using the verbiage found in a recent USAF Doctrine Center publication, *50 Questions Every Airman Can Answer* that uses an adjectival form of the term:

> Aerospace power is essentially the ability to create political and military effects using aircraft, spacecraft, and information. Aerospace power involves the effective use of the full range of the nation's resources to allow us to use the physical environments of air and space and our information resources to our national advantage. Air Force Doctrine Document 1, *Air Force Basic Doctrine,* defines the combination of air and space power as the synergistic application of air, space, and information systems to project strategic military power.[20]

This description presents the notion that aerospace power is a philosophy built upon three pillars; airpower, spacepower, and information power. It therefore becomes necessary to contemplate each form of power on its own merits.

Airpower. It is best to remain consistent with the description of airpower provided in *50 Questions Every Airman Can Answer,* but notice how well the functional definition of air fits the following:

> Airpower is the fundamental ability to use aircraft to create military and political effects. Another way of defining it is 'military power that maneuvers through the air while performing its mission.' Airpower is a subset of aerospace power.[21]

---

[18] AFDD 2, *Organization and Employment of Aerospace Power,* 17 February 2000, 133.

[19] The maritime environment is defined as "the oceans, seas, bays, estuaries, islands, coastal areas, and the airspace above these, including amphibious objective areas." Found in Joint Pub 1-02 *Definitions Related to Command and Control,* 14 June 2000, 280.

[20] Baier, 3.

[21] Ibid., 6.

Spacepower. (This is a definition of military spacepower only, not the broader definition promulgated above.) Here again, the functional definition of space fits well into the following description of spacepower from *50 Questions Every Airman Can Answer:*

> Much like airpower, space power is, in essence, the ability to use spacecraft to create military and political effects. Another way of saying it is 'military power that comes from, resides in, or moves through space while performing its mission.' Space power, like airpower, can place an adversary in a position of disadvantage. Space power is a subset of aerospace power.[22]

The Evidentiary Base

Books, articles, reports, studies, doctrine, policy letters, and personal interviews form the evidentiary base for this study. Two criteria are required of the evidence used to construct and argue propositions regarding spacepower. First, to the extent possible, evidence for propositions must be rooted in experience, not untested theory. Second, the evidence must be stripped of bureaucratic and organizational prejudice. Given these criteria, it is clear that more recent source materials are preferred since the modern experience base is broader. However, the majority of recent works are advocacy pieces that all but ignore spacepower while arguing—sometimes quite passionately—for a preferred organizational model in an attempt to manage space systems more effectively. While this limits their contribution to framing a spacepower proposition per se, such works are nonetheless valuable because they provide excellent arguments used to support or attack certain propositions presented in this study.

---

[22] Ibid., 7.

The origins of American spacepower are unique when compared to other forms of military power. Whereas land, sea, and air power evolved out of private and commercial endeavors, spacepower did not. Moreover, the other forms of military power expressed themselves fully in the First and Second World Wars—arguably total wars fought without much restraint. Spacepower did not because it had not yet arrived on the scene. This difference affects the evidentiary base for any study regarding spacepower.

The story of US military spacepower begins in the mid-1940s with notes from General Hap Arnold to Dr Theodore von Karman inquiring about the untapped potential of space. This was in the wake of the Second World War, after the use of the atomic bomb, in the fledgling days of the United Nations, and on the eve of the Cold War. Consequently, space systems developed under the strictest military secrecy and with considerable presidential oversight as large governmental endeavors. For these reasons, some of the most basic space-based capabilities remained a mystery to the public and much of the military until recently. This stands in stark contrast to the developments of land, sea, and air power. Perhaps this explains why no great spacepower theory has been forthcoming despite more than fifty-five years of contemplation. In its place, civilian authors created an entire literary genre of space-related science fiction and science fantasy. Television shows and movies such *as Star Trek, 2001: A Space Odyssey, Star Wars, Battlestar Gallactica, Alien,* and *Starship Troopers,* to name just a few, have permeated the popular culture and planted fantastic and quite unrealistic ideas about space. Part of the struggle to make America and its military members more aware of the true nature of spacepower requires undoing what the media has done.

While there is no great book about spacepower, recent years have witnessed an exponential growth in the body of papers, articles, speeches, and other documentation focused on American spacepower. Military members fulfilling course requirements during professional military education generate many of these works, while private think tanks, such as the RAND Corporation, Congressional studies, and a handful of interested civilian authors contribute with growing frequency.

Three occurrences in the last twenty years prompted this explosion of critical thought and publications regarding spacepower. The first was President Reagan's National Missile Defense or "Star Wars" proposal, which generated international debate, but failed to materialize. The second is the ever-increasing military and commercial reliance on space systems, which now form a significant national infrastructure requiring protection. Finally, the ongoing effort to transform the American military in the post-Cold War era places emphasis on exploiting new technical capabilities such as those offered by space systems—along with a revised proposal for a missile defense system by President George W. Bush.

America has pursued space operations for several decades, but the nation at large is only now realizing the great-untapped potential of space. There is little doubt that the evidence supporting propositions regarding spacepower is still ripening, but the evidentiary base is nonetheless sufficient from which to draw reasonable propositions regarding the fundamental nature of spacepower.

Methodology

The method used to derive the propositions presented here involved three years of literary research and extensive personal interviews that resulted in a long list of statements regarding spacepower. The list evolved during numerous brainstorming sessions with several space experts—most of them space weapons officers with combat experience.[23] During more than thirty-six months of debate, the list was carefully refined into the ten most salient propositions. These are the ten propositions regarding spacepower. In sifting through the evidence, arguments supporting and refuting each proposition emerged. The arguments on both sides appear with each corresponding proposition in chapter 3.

At the turn of the twenty-first century, we do not have experience fighting wars in, from, and through space, in the classic sense. Therefore, many people view military space activities as merely an avenue to support the information needs of terrestrial forces. While this is certainly important, spacepower is much more than support, as the propositions point out. While the propositions are rooted in space experience to date, it is proper to use analogies to other forms of power to predict, within reason, certain ways spacepower is likely to evolve. The case in point is the last proposition, that the weaponization of space is inevitable. Support for this proposition comes from the historical evidence that shows that humans have always weaponized the different media. Therefore, it is reasonable to predict that weaponizing space will also occur.

Analytical Criteria

To answer the central question of this study, "what is the nature of spacepower?" the evidence must culminate in propositions that describe various aspects of spacepower. In turn, each proposition must serve as a logical answer to the central question. Each

---

[23] A space weapons officer is a space operations officer (13SXX) graduate of the Space Weapons Instructor Course at the USAF Weapons School at Nellis AFB, NV. This is the USAF's graduate-level training program for the tactical and operational employment of American space systems in combat.

proposition must also serve as a premise for an argument supporting or refuting the hypothesis presented below.

A proposition regarding spacepower may be very similar to a valid proposition regarding some other form of military power. This does not invalidate a claim that spacepower is different. In fact, similarities should occur for the same reasons that the principles of warfare are common to all forms of military power. As Clausewitz described war, "its grammar, indeed, may be its own, but not its logic."[24] If military spacepower is truly a form of power, then it should fall in line with a common logic guiding the reasoning of other forms of warfare. As such, similarities to propositions that apply to other forms of military power are expected.

Hypothesis

The hypothesis tested in this study is the assertion of the aerospace integration school, namely that spacepower is simply a continuation or extension of airpower; that it is not an independent form of power. The test of this hypothesis occurs in the presentation of evidence supporting and refuting each proposition presented in chapter 3 (Ten Propositions Regarding Spacepower). Finally, this hypothesis is evaluated in the concluding chapter of this study after presenting and debating all of the evidence.

Limitations of this Study

When Meilinger framed his propositions regarding airpower, he had more than eighty years of airpower history to draw from. His overwhelming evidentiary base included two world wars and hundreds of other air campaigns among dozens of nations.

In the wake of Operations Desert Storm and Allied Force, many now claim that airpower has become the force that "can do most of the work" in modern combat.[25] Most nations maintain separate air forces and many have academic institutions dedicated to the study of airpower as a distinct discipline. Spacepower advocates can make no such claims, although Russia recently established an independent space force. It remains unclear whether Russia's initiative is a sign of support for an independent role for spacepower or indicative of some systemic organizational weakness.[26]

The discussion of the evidentiary base of this study points to a stark limitation; the evidentiary base is meager in comparison to that available to air, land, and sea advocates. Spacepower is different from other forms of military power because it came into being after World War II and evolved under a shroud of secrecy during a highly politicized era of limited warfare. It is unreasonable to expect our national concept of spacepower to match that of air, land, and sea power since it has yet to be pushed to its extreme by total warfare. It also lacks the romantic heraldry of battle-proven warfighters passing the torch to succeeding generations. The lack of glamour undoubtedly dissuaded some would be authors from contributing to serious spacepower literature.

When conducting personal interviews there arises a strange dichotomy. The more senior ranking the interviewee (colonels and higher), the less likely he or she has any operational experience with space systems. However, they are increasingly likely to be involved in the bureaucratic and organizational policy making affecting military-related

---

[24] Carl von Clausewitz, *On War*, ed. and trans. Michael Howard and Peter Paret (Princeton N.J.: Princeton University Press, 1984), 605.
[25] Robert A. Pape, *Bombing to Win: Air Power and Coercion in War*, (Ithica, N.Y.: Cornell University Press, 1995), 326, and Benjamin S. Lambeth, *The Transformation of American Airpower*, (Ithica, N.Y.: Cornell University Press, 2000), 7.
[26] Nikolai Novichkov, "Russia Gives a Boost to Space Effort," *Jane's Defense Weekly*, 7 Feb 2001.

space operations. Conversely, the more junior the interviewee (captain to lieutenant colonel), the more likely he or she has extensive experience operating space systems, but with little insight into the jungle of politics surrounding their senior decision-makers. This is because space operators have not progressed to the senior ranks inside the Air Force within their operational career field. Because of this dichotomy, special care is required when treating the information passed along by the interviewee in order to meet the standards established for the evidentiary base (experientially based, and stripped of bureaucratic and organizational politics).

Another limitation of the evidentiary base is the lack of extensive human experience in the space environment itself. Other forms of military power are much more tactile in nature, since humans physically operate their equipment in the other media. One can easily imagine the bygone days when some dashing aviator climbed out of his Sopwith Camel on some misty French aerodrome and manually tightened the guy wire supporting one of his wings. This image is literally lost in space. Military spacepower predominantly employs unmanned robots in the space environment. Invisible data streams carrying ones and zeros between operator and satellite, and from satellite to user, does not conjure a romantic image. Nor does it provide a concrete image. Yet, this unique form of military power is of ever increasing importance in modern warfare.[27] Between wars, airmen raised the "airmindedness" of the nation by barnstorming and continuing this tradition with airshows. Space has no such opportunity, with the exception of the enthusiasm for space exploration created by the National Aeronautics and Space Administration (NASA) by broadcasting images and video from the Space Shuttle, the International Space Station, and numerous exploratory probes. Various space camps around the country also indoctrinate youths into the exciting aspect of manned spacepower, but it does not create a general awareness of military spacepower, particularly regarding the use of uninhabited satellites. Space professionals perceive this lack of "spacemindedness" on the behalf of the public and fellow service members. Consequently, authors of spacepower documents often limit their works to very elementary discussions. This is pervasive across the genre of spacepower literature. In framing propositions, care is given to keep the language simple, while encompassing the expert-level issues faced by the space professionals themselves.

Some critics will argue that it is premature to suggest propositions regarding spacepower. Their argument has merit only if we trivialize our present operational experience in space. The evidentiary base is comparatively meager, admittedly, but it is no less important than understanding the foundations of any other form of power. There is no attempt to raise these propositions to the level of tenets of spacepower, although that may eventually happen. It is, however, vital to the continued growth of America as a spacepower nation that basic propositions regarding spacepower be laid down to promote deeper understanding of this increasingly important medium. Critics must answer the following question which has been asked repeatedly in several top level commissions, "if not now, when?"

Overview

Chapter 2 (Schools of Thought in American Spacepower) provides a brief summary of the evolution of American spacepower doctrines using the framework proposed by David E. Lupton in his work *On Space Warfare*. The purpose of this discussion is to

---

[27] Maj Gen John Barry, lecture, School of Advanced Airpower Studies, Maxwell AFB, Ala., 3 April 2001.

familiarize the reader with the political and military environment in which spacepower evolved and operates. It exposes shifts in the geopolitical landscape that may cause America to dramatically alter its spacepower doctrine, making it more important than ever for policy makers to understand the fundamentals of spacepower.

Chapter 3 (Ten Propositions Regarding Spacepower) is the core of this work. Presentation and argument of the ten propositions regarding spacepower occurs here. The purpose of this chapter is to deliver the evidence addressing the central question of this study; "what is the nature of spacepower?" This evidence also supports the testing of the hypothesis that spacepower is merely a continuation or extension of airpower; that it is not an independent form of power.

Chapter 4 (A Spacepower Theory) answers the central question of this study by describing the nature of spacepower as revealed in the evidence for the ten propositions. It is here that the hypothesis is specifically accepted or rejected. During the fact-finding phase of this study there were some significant lessons learned about spacepower that appear here. Finally, and most importantly, with an understanding of the nature of spacepower, this study concludes with a brief spacepower theory.

The Appendixes provide a brief look at the attempts of other authors to frame statements that capture the essence of spacepower. In a sense, these are their propositions regarding spacepower, but this was not the focus of their works. The purpose of reviewing the work of other authors is to provide readers with additional information from which to judge the propositions regarding spacepower that are central to the work at hand. In some cases, the propositions presented here overlap with the work of other authors, presumably because an inherent truth is apparent. In other cases, the propositions are dissimilar to the work of other authors because of differing intents, methodologies, analytical criteria, and focus of their works. The recommendations of the Space Commission Report also appear in an appendix because this landmark document will likely influence future spacepower debates and policy-making.

Chapter 2

Schools of Thought in American Spacepower

History, by apprizing [men] of the past, will enable them to judge of the future; it will avail them of the experience of other times and other nations; it will qualify them as judges of the actions and designs of men; it will enable them to know ambition under every disguise it may assume; and knowing it, to defeat its views.

—Thomas Jefferson, Notes on the State of Virginia, 1784.

Contemporary critics of spacepower have too little sense of history. Whatever wonders "the stars" hold for our future, there is a vastly nearer-term strategic logic of spacepower that is all but entirely comprehensible in principle today.

—Dr. Colin S. Gray

"Because it's there," was the infamous reason Sir George Leigh Mallory gave for wanting to climb Mount Everest. Some may argue this is why America went to the moon (if they dismiss the perceived race against the Soviets and the amount of international prestige [read as power] associated with being the first). The reasons for exploiting the space medium for military advantage are far more practical than the simple quest for adventure, knowledge, or glory.

The reasons for pursuing spacepower are perhaps best summarized by Thucydides, who explained more than two-thousand years ago that *"fear, honor, and interest"* serve as the three strongest motives for taking action.[28] These three motives help explain much of

---

[28] Thucydides, *The Peloponnesian War*, ed. Robert B. Strassler, *A Comprehehsive Guide to the Peloponnesian War* (New York: Free Press, 1996), 43. Thucydides is recognized as the spiritual father of realism. The realist perspective focuses on the state as a unitary actor in an anarchic international system. Realists argue that states pursue security, autonomy, and power relative to one another, and that conflict is caused by security dilemmas wherein a state fears an unacceptable loss of power forcing that state (or

the underpinnings of America's military space effort. This chapter surveys how these motives have driven the evolution of American spacepower using the David Lupton's framework.[29]

Lupton's Four Doctrines

In his 1988 work, *On Space Warfare: A Space Power Doctrine*, author David E. Lupton provides a comprehensive framework to analyze the rationale for various military activities in space. He describes four main schools of thought associated with military spacepower. He calls them the sanctuary, survivability, control, and high-ground doctrines. They represent an escalating spectrum of commitment to military spacepower as a source of national and military power.

The Sanctuary School

> A fundamental tenet of this school is that the primary value of space forces is their capability to "see" within the boundaries of sovereign states. This value stems from the space vehicle's legal overflight characteristic. Proponents of sanctuary doctrine argue that past arms limitations treaties could not have been consummated without space systems that serve as a "national technical means of treaty verification." Moreover, the prospects for any future treaties would be extremely dim without the ability of space systems to fulfill President Eisenhower's dream of [treaty] verification through open skies. Thus, space systems have had a tremendous stabilizing influence on relations between [states]. Finally, these advocates caution that overflight is a granted right that nations have not attempted to deny and that any proposed military use of space must be weighed against the possible loss of peaceful overflight. This train of thought leads to the conclusion that the only way to maintain the legal overflight characteristic is to designate space as a war-free sanctuary.[30]

---

states) to act. Discussed in Barry B. Hughes, *Continuity and Change in World Politics, 3rd ed.* (Upper Saddle River, N.J.: Prentice Hall, 1997), 46-53.
[29] David E. Lupton, On Space Warfare: A Space Power Doctrine (Maxwell AFB, Ala.: Air University Press, June 1988)
[30] Ibid., 35.

The basic tenet of the sanctuary school is that satellite reconnaissance systems made nuclear war less likely and fostered stability in the superpower relationship.[31] From a realist's perspective, exploiting overflight rights for satellite reconnaissance of potential adversaries reduces potential security dilemmas brought about by *fear* of the unknown. In one sense, it is a vaccine against a surprise attack, like Pearl Harbor. With the knowledge gained from space, policy makers are better able to assess their national security situation and scale their defense expenditures appropriately. This makes other resources available for a state to pursue its *interests* elsewhere. At the dawn of the space age, satellites were a symbol of great national achievement. This built an aura of prestige or *honor* associated with space programs which the superpowers used to attract third-world nations to their causes, thereby gaining power. Sanctuary school advocates seek to preserve space as a weapons-free zone to prevent threatening other states and triggering security dilemmas. Weaponizing space may not only diminish the honor (power) of the offending state, but may also entice other actors to contest the established rights of international overflight by satellites. If successful, this could culminate in closing off reconnaissance access to many parts of the geopolitical world—hence returning the space powers to security dilemmas rooted in fear of the unknown.

The Survivability School

> The basic tenet of this school is that space systems are inherently less survivable than terrestrial forces. Several factors undergird this belief. First are the long-range weapon effects in the space environment, coupled with a belief that nuclear weapons are more likely to be used in the remoteness of space. Second, the quasi-positional nature of space forces and their vehicular sovereignty imply that space forces cannot rely on maneuverability or terrestrial barriers to increase survivability. Finally, the negative aspect of space forces' political insensitivity creates

---

[31] Lt Col Peter L. Hays, et al., eds., *Spacepower for a New Millennium* (New York: McGraw-Hill, 2000), 3.

uncertainty about the political implications of an attack on space forces (e.g., would we go to war if a satellite were destroyed?). Advocates of the survivability school have serious reservations as to the military value of space forces. They agree that space forces can do certain military functions (e.g., communication and weather data gathering) more economically and efficiently in peacetime than other forces. They believe, however, that space forces must not be depended on for these functions in wartime because they will not survive.[32]

Advocates of the survivability school concede that space is an excellent place to base many military systems, especially those that augment or enhance terrestrial forces. Realists point out that it is in a state's *interest* to exploit space to gain an advantage in military power over other states. Triggering a security dilemma by developing space-based force enhancement capabilities is not particularly worrisome. However, exposing vulnerabilities to an enemy by virtue of relying on "fragile" space systems is a constant source of *fear* by the state that uses space assets for military and economic purposes.

The Control School

The control school declines to place an exact value on space forces and only suggests their value by using air power and sea power analogies. For example, according to Gen Thomas A. White, "Whoever has the capacity to control space will likewise possess the capacity to exert control over the surface of the Earth." Others argue that there are space lanes of communications like sea-lanes of communications that must be controlled if a war is to be won in the terrestrial theaters. Control school advocates argue that the capability to deter war is enhanced by the ability to control space and that, in future wars, space control will be coequal with air and sea control.[33]

Control school advocates believe space is just another medium analogous to air, land, and sea. As such, they advocate controlling this medium vigorously with both offensive and defensive operations. They believe space control is essential to securing victory in any terrestrial conflict. A realist would point out that it is in a state's *interest* to

---

[32] Lupton, 36.

control space because it is a route of commerce comparable to terrestrial lines of communication. *Fear* of triggering security dilemmas in other states is not a dominant concern in this school of thought.

The High Ground School

> [This] school harkens back to the old military axiom that domination of the high ground ensures domination of the lower lying areas. Disciples of this "high-ground" school advocate a space-based ballistic missile defense (BMD). They argue that the global presence characteristic of space forces combined with either directed-energy or high-velocity-impact space weapons provide opportunities for radical new national strategies. In their view, space-based defensive forces can reverse the current stalemate caused by the preeminence of the offense and create either an offensive-defensive balance or a preferred defensive stalemate. This rebalancing would allow replacement of the flawed strategy of assured destruction with one of assured survival. The high-ground school believes space forces will have a dominant influence.[34]

"Lupton's final doctrine, high-ground, argues that space is the dominant theater of military operations and is capable of affecting terrestrial conflict in decisive ways." [35] This is analogous to Douhet's implicit contention that aircraft are the solution to strategic and tactical stalemates and that all future wars can be won from the air.[36] High ground advocates favor full weaponization of space, featuring missile defense systems. Space control is essentially a prerequisite. A realist would contend that the high-ground school is the ultimate exploitation of space to secure the *interests* of the state and to avert *fear* of another state exploiting space against those interests. Like so many airpower theorists in the early days of aviation, high-ground advocates rest their case regarding the decisive

---

[33] Lupton, 37.
[34] Lupton, 36-37.
[35] Hays et al., 3-4.
[36] Giulio Douhet, *The Command of the Air*, trans. Dino Ferrari (Washington, D.C.: Air Force Museums and History Program, 1998), 15-29.

24

nature of space upon optimistic speculation with little empirical support. Figure 1

summarizes Lupton's four military space doctrines.

Figure 1. Attributes of Military Space Doctrines

	Primary Value and Functions of Military Space Forces	**Space System Characteristics and Employment Strategies**	**Conflict Missions of Space Forces**	**Desired Military Organizations for Operations and Advocacy**
Sanctuary	• Enhance Strategic Stability • Facilitate Arms Control	• Limited Numbers • Fragile Systems • Vulnerable Orbits • Optimize for NTMV	• Limited	None
Survivability	Above functions plus: • Force Enhancement	• Redundancy • Hardening	• Force Enhancement • Degrade Gracefully	Major Command or Unified Command
Control	• Control Space • Significant Force Enhancement	• On-Orbit Spares • Crosslinks • Maneuver • Less Vulnerable Orbits • Stealth • Reconstitution Capability • Defense • Convoy	• Control Space • Significant Force Enhancement • Surveillance, Offensive, and Defensive Counterspace	Unified Command or Space Force
High Ground	Above functions plus: • Decisive Impact on Terrestrial Conflict • BMD		Above functions plus: • Decisive Space-to-Space and Space-to-Earth Force Application • BMD	Space Force

Source: Hays, et al., 4.

The Historical Evidence

Since the dawn of the space age, there have been advocates for each of Lupton's schools of thought. In general, it is fair to say that America followed the sanctuary doctrine during the Cold War. After the Cold War drew to a close, America began focusing on the force enhancement opportunities offered by space-based systems, signaling a move towards a survivability doctrine. General Eberhart, the Commander-in-Chief of US Space Command (USCINCSPACE), is now leading the charge in advocating that America pursue a new doctrine—"space control."[37] The remainder of this chapter will survey the evolution of spacepower doctrine using Lupton's model and will look at the current conditions that may lead to a shift in that doctrine. It is necessary to ask what has changed, and is a move towards a space control strategy feasible in the current political and military context. Answering these questions is important because adopting a space control strategy would be a significant departure from long-standing space policy.

Sanctuary Doctrine (Cold War to Desert Storm)

Just as military aviation was born in the First World War as a method of conducting reconnaissance over enemy territory, so too were military satellites born to conduct reconnaissance over enemy territory during the Cold War. Arguably, the greatest security dilemma facing the US in the 1950s was the fear of a nuclear showdown with the Soviet Union. This drove the need for gathering intelligence about the Soviet Union—a closed society. Accordingly, in May 1955 the Eisenhower administration established its intent to launch a satellite "to establish a legal regime to legitimize overflight and thereby

---

[37] Space control refers to the use of force to gain access to space while denying the same to an adversary. It is a relative term and does not imply absolute control. It may include using weapons placed in space such as missile defenses.

open up the closed Soviet state to satellite reconnaissance by the secret WS-117L spysat system. This policy, along with Eisenhower's [concern over the growing power of the military-industrial complex], also led to the creation of the National Reconnaissance Office (NRO), America's secret and independent military space agency."[38] Establishing the NRO also provided a convenient civilian cover story for the building and launching of reconnaissance satellites.

Eisenhower succeeded in making unimpeded satellite overflight acceptable to the international community—a task made easier by the Soviets who were the first to establish the precedent with the launching of *Sputnik* on 4 October 1957. Subsequent administrations would formally hammer out the corresponding treaties in the United Nations that set aside space for peaceful purposes only and preventing any nation from claiming sovereignty over any part of space.[39] In effect, these efforts established space as a "sanctuary" for peaceful operations. These principles are now a matter of accepted international law and continue to influence the US National Space Policy and military space strategy in very fundamental ways.

Until Desert Storm in 1991, space systems were essentially an adjunct to other political and military operations. At that time the Air Force mission statement read: "To defend the United States though the control and exploitation of the air."[40] Any reference to space was missing because despite more than thirty years of Air Force space activity, the capabilities of space systems were unknown to the Air Force at large, and what

---

[38] Lt Col Peter Hays and Dr Karl Mueller, "Boldly Going—Where?" *Aerospace Power Journal* 15, no. 1, (Spring, 2001), 36.
[39] These principles are laid out in the Outer Space Treaty of 1967.
[40] R. Cargill Hall and Jacob Neufeld, *The US Air Force in Space: 1945 to the Twenty-First Century* (USAF History and Museums Program, United States Air Force, Washington D.C., 1998), 174.

existed was limited to a supporting role. This was mostly because space systems evolved to support the extremely sensitive nuclear mission of the Strategic Air Command (SAC), which dominated the Air Force during the Cold War. Consequently, the Air Force space program focused on providing missile attack warning, global weather, global positioning, and global command and control communications to support the National Command Authority (NCA) and SAC. Likewise, reconnaissance support from the NRO was limited to the NCA, SAC, and very few others.

The sanctuary doctrine was highly effective in this era for a number of reasons. First, the National Space Policy and military space strategy were perfectly aligned. Both gave top priority to supporting the nuclear mission. Second, the US had only one peer competitor, the Soviet Union, which was also interested in establishing its own spysat network. This made it easy for diplomats to secure the principles discussed above. Third, although the Soviets and US pursued some space control capabilities such as anti-satellite (ASAT) technologies, none matured to the point of presenting significant threats to space capabilities. Ultimately, the US cancelled all of its ASAT programs because of technical difficulties and a lack of political will to commit strained budgetary resources in the face of marginal threats. Cancellation also alleviated concerns about weaponizing space in violation of the space sanctuary doctrine. Finally, given the limitations of satellites themselves, computing power, and the extremely high classification of space products during this period, integrating space capabilities into operations at levels lower than the NCA and strategic planning centers was rare and difficult at best. In reality, technical limitations may have constrained America to follow the sanctuary doctrine.

Survivability Doctrine (Desert Storm to the Present)

Desert Storm was a watershed event for spacepower.[41] The Cold War was over and the fear of a surprise nuclear attack withered along with the emphasis on nuclear deterrence. This allowed space operators to strengthen their emphasis on providing force enhancements to conventional warfare, as described by Lupton's survivability doctrine. Desert Storm gave them an opportunity to showcase their capabilities. During the war, missile warning satellites and the Global Positioning System (GPS) became the sweethearts of the USAF space program. Unfortunately, the warfighters typically lacked the appropriate security clearance, the proper ground exploitation tools, and the training to use other types of intelligence-related space support available to senior political and military leaders. In the aftermath of the war, many generals complained bitterly about these shortcomings.[42] However, most agreed that spacepower was a significant contributor to the war effort, but much more was possible.

In June of 1992, General Merrill McPeak, the Chief of Staff of the Air Force, gave a speech at Maxwell Air Force Base in which he announced a change to the Air Force mission statement. He added the words "air *and space*."[43] This represented an overt institutional shift toward policies inherent in Lupton's survivability doctrine. It essentially entails the Air Force and other Services working to provide force

---

[41] Desert Storm is often cited as "the first space war." This is only true if you discount its contributions to the Cold War and the Vietnam War. One can argue that Desert Storm was the first time that space operations effected every American in the battlespace. My favorite quote comes from a young Marine who told a CNN reporter, "Space didn't have anything to do with Desert Storm. All I needed was my M-16 and this little box that tells me where I am (pointing to his Global Positioning System (GPS) receiver)!" From the 57th Training Support Squadron, "Global Positioning System Overview," USAF Space Weapons Instructor Course, MSN574Z, 2 February 1999.

[42] This was General Schwartzkopf's number one complaint. General "Chuck" Horner was the Joint Forces Air Component Commander (JFACC) during the war. He later became USCINCSPACE and frequently related his frustration at trying to integrate space into the fight only to be thwarted by security barriers.

[43] Hall and Neufeld, 174.

enhancements by pushing space products down to the operational and tactical users by removing security barriers, providing training, and acquiring newly available computers powerful enough to exploit space derived data.[44] It is essentially the integration of space capabilities with terrestrial forces to enhance terrestrial operations.

The shift toward the survivability doctrine was consistent with White House policy as evidenced by President Bush's decision to declassify the existence of the NRO in 1992, making access to space derived data much easier. From a fiscal perspective, space integration also made sense since it created opportunities to eliminate unnecessary duplication of effort, such as replacing some aircraft reconnaissance platforms with satellites and ground radio navigation aids with GPS. At the same time, it brought new capabilities to the campaign planners and warfighters. More importantly, adopting the force enhancement elements of the survivability doctrine did not cross the imaginary line in the sand of weaponizing space, therefore it did not set off significant security dilemmas amongst other states.

The integration of space-derived force enhancements into terrestrial operations is succeeding in the opinion of General Eberhart, who recently commented, "The fact we heard so much about [the need for integration] after Desert Storm, and didn't after Kosovo, tells me we're on the right track."[45] Regardless of what other spacepower doctrines the US might pursue, it appears that force enhancement will continue as more and more systems are plugging into space derived information sources. In fact, at the turn of the century there is a growing tendency to exploit commercial space assets to support military operations.

---

[44] Ibid.

Control Doctrine (Future?)

In November 2000, General Eberhart stated, "I don't think we would be good stewards of space capabilities if we only thought about 'integration.' We also need to be spending resources and intellectual capacity on space control."[46] In essence, the general joins control doctrine advocates who favor developing the capability to use force (when required) to secure American access to space and to deny the same to an adversary. Many of these same advocates also support President George W. Bush's call for a missile defense system. This represents a significant departure from the doctrinal philosophy of either the sanctuary or the survivability schools, as it may entail negating adversary satellites on orbit and crossing the threshold of putting weapons in space—in line with the high-ground school of thought.

What has changed? First, the continuing force enhancement effort makes all of America's armed forces increasingly reliant on space support. This creates vulnerabilities predicted by the survivability doctrine that require protection. Second, commerce in the Western World has also become increasingly reliant on satellites for the collection and routing of essential information. The international banking community, the global telecommunications industry, and the stock markets became heavy users of space services in the 1990s. It is unclear whether or not the commercial sector understands the threat, or wants protection.[47] Third, most other nations are also increasing their use of space. Many of these nations are potential adversaries who may

---

[45]William B. Scott, "CINCSPACE: Focus More on Space Control," *Aviation Week & Space Technology*, 13 November 2000, 80-81.

[46] Ibid., 80.

[47] Commercial satellite operators typically do not want to add protection to their satellites because doing so would increase weight and therefore their launch costs would go up. During the Schriever 2000 wargames in Colorado Springs one corporate leader of a satellite company told Maj Gen Lance Smith, "Protection?

31

exploit their space capabilities against the US if they believe they will benefit by doing so. Finally, several countries, such as Russia and China, have already developed counter-space weapons that directly threaten US space operations.[48] Other states are likely to proliferate these or similar weapons in the coming decades. In sum, the US military increasingly relies upon space assets, therefore it is in their interest to protection them from enemy attack. At the same time, the US has an interest in denying adversaries the ability to exploit their own space assets to gain an advantage. *In light of these changes, General Eberhart's assertion that America needs to focus more on space control may be warranted from a military strategist's point of view, but is it prudent from a national policy perspective?*

The latest National Space Policy (19 Sep 96) is contradictory and confusing regarding space control. It asserts, "The US is committed to the…use of outer space by all nations for peaceful purposes," and "considers the space systems of any nation [to have] the right of passage through and operations in space without interference."[49] This is consistent with the sanctuary doctrine. However, the document later asserts that the "[Department of Defense] DoD shall maintain the capability to execute the mission areas of space support, force enhancement, *space control*, and force application."[50] The National Space Policy advocates both sanctuary and control doctrines, but only one or the other doctrine is practicable.

That's what insurance is for." Maj Gen Lance L. Smith, Maxwell AFB, Ala., interview with author, 9 March 2001.

[48] Gen Richard B. Meyers, "Space Superiority is Fleeting," *Aviation Week & Space Technology*, 1 January 2000, 54.

[49] *National Space Policy*, The White House, 19 September 1996, 1.

[50] Ibid., 4 (Emphasis added).

Actions speak louder than words. The DoD currently has little or no space control capability. Despite the initiation of various space control programs by the Kennedy, Ford, and Reagan administrations, subsequent administrations cut these programs in favor of abiding by the sanctuary doctrine. For example, the Clinton administration cut, cancelled, or delayed several space control related initiatives such as Clementine II, the Space Plane, and the Kinetic Energy Anti-Satellite Weapon. He also deferred the approval for National Missile Defense (NMD) component construction (mainly radars) to his successor—a system Lupton describes as part of the high ground school of thought.

The appeal of the sanctuary doctrine to political leaders is rooted in four basic assumptions. First, since the US is the most dependent on space support for its economic and military interests, it therefore has the most to lose by abandoning the sanctuary doctrine. Second, pursuing a space control doctrine might destabilize the world balance of power since it threatens space assets used by other nations in national security roles, thereby causing security dilemmas of an unpredictable nature.[51] Third, during the Clinton administration, several intelligence agencies argued that there were no operational threats to American space systems, therefore there was no need to make the heavy financial investments to fulfill a space control strategy. To the contrary, in addition to long standing Russian ASAT programs, the Chinese openly declared that they have an operational ground-based ASAT laser.[52] Furthermore, the Chinese are also

---

[51] John B. Sheldon, "Space as the Forth Environment: For Warfare or a Supporting Role?" *RUSI Journal*, London, October 1999, 2.

[52] "China Develops Anti-Satellite Laser System," *Jane's Defense Weekly*, 2 December, 1998, 18. The Russian and Chinese ASAT programs are further elaborated on by Vice Adm. Thomas R. Wilson, "Global Threats and Challenges Through 2015," (Statement for the Record Senate Select Committee on Intelligence, 7 February 2001), 14, on-line, Internet 12 April 2001, available from http://www.dialumni.org/images/dr_testimony.pdf.

developing "parasitic satellites" for use as ASATs.[53] Finally, and perhaps most importantly, the US may lose diplomatic power by abandoning the sanctuary doctrine. NATO allies have repeatedly voiced concerns regarding America's propensity to push technology beyond their ability to integrate as equal partners in the alliance, with fears of the US destabilizing the global power structure. Moreover, the UN Secretary General, Kofi Annan, is a strong advocate of the sanctuary argument and believes that "the advantages of space technologies should be shared amongst everyone, and military conflict in space threatens this prospect."[54] *It appears the desire to shift towards a space control doctrine was not possible during the Clinton administration for political reasons, given the administration's policy of following the sanctuary doctrine.*

Despite General Eberhart's position as USCINCSPACE, he was unable to significantly influence the National Space Policy during the Clinton administration toward a space control doctrine because he was only one voice in the larger strategic culture of space policy decision-making. When General Charles Horner was USCINCSPACE in the mid-1990s, the chairman of the Senate Armed Services Committee, Senator Sam Nunn, asked him if he was in charge of space. General Horner replied that it depends because he cannot exercise unitary control over his own command. USCINCSPACE currently shares the responsibility for making military space policy with several other civil and military agencies as shown in Figure 2.

---

[53] Cheng Ho, "China Eyes Anti-Satellite System," *Space Daily*, 8 January, 2000.
[54] Kofi Annan, "Keep Space Peaceful," *Space News*, 23 August 1999, 4.

Figure 2. Major US Agencies Involved in Space Policy

- **Executive Office of the President**
  - Office of the Vice President
  - Office of Science and Technology Policy
  - Office of Management and Budget Office of the U.S. Trade
  - Representative National Science and Technology Council
  - National Security Council and the Space Policy Coordination Committee
- **National Aeronautics and Space Administration (NASA)**
- **Department of Defense (and related agencies)**
  - US Space Command
  - Army, Navy, Air Force, and Marine Corps
  - National Reconnaissance Office
  - National Imagery and Mapping Agency
  - National Security Agency
  - Defense Advanced Research Projects Office
  - Defense Information Systems Agency
  - Ballistic Missile Defense Office
- **Department of Commerce**
  - National Oceanic and Atmospheric Administration (NOAA)
- **Department of Transportation**
  - Office of Commercial Space Transportation, Federal Aviation Administration
- **Department of State**
  - Office of Defense Trade Controls
- **Central Intelligence Agency**
- **Federal Communications Commission (FCC)**

Source: Frank G. Klotz, *Space, Commerce, and National Security* (New York: Council on Foreign Relations Press, 1998), on-line, Internet 12 May 2001, available from http://www.foreignrelations.org/p/pubs/klotz.html.

These agencies intrude on the USSPACECOM budget, resources, and decision-making authority. "In addition to the governmental intrusion into his joint command, USCINCSPACE must also deal with service infighting over who should have the dominant role in space."[55] The Air Force believes it should play the dominant role since it dedicates more resources to space missions than the other services. It is important to

---

[55] Bruce M. DeBlois, *Beyond the Paths of Heaven* (Maxwell AFB Ala.: Air University Press, September 1999), xiii.

note that the Air Force embraced the survivability doctrine, pressing the space community to provide greater force enhancement capabilities to airmen. In its *Global Engagement* vision of November 1996, during General Ronald Fogleman's tenure as Chief of Staff, the Air Force issued what is probably the most strident position ever regarding the importance of space to the Air Force's future: "We are now transitioning from an *air and space* force on an evolutionary path to a *space and air* force."[56] The enthusiasm was toned down considerably in 1998 by the subsequent Chief of Staff, General Michael Ryan, who insisted that air and space form a single, seamless aerospace medium. This is the basis for the Air Force's "aerospace integration" effort, which attempts to blend the air and space communities into a single body of likeminded professionals.

From a bureaucratic politics perspective, all outward appearances of this effort suggest it is merely an attempt by the Air Force to secure future missions and resources for itself over its sister services.[57] Only the Air Force views space as part of an "aerospace continuum." This represents a major departure from the current National Space Policy and Department of Defense Space Policy, not to mention the Annual Report to the President and the Congress by the Secretary of Defense (2000). All of these documents refer to space as a separate medium with spacepower being of equal

---

[56] Gen Ronald R. Fogleman and The Honorable Sheila E. Widnall, *Global Engagement: A Vision for the 21st Century Air Force* (Washington, D.C.: Department of the Air Force, November 1996), 8. Emphasis is original.
[57] This was asserted during the hearings on "Missile Development and Space Sciences" in 1959 during the testimony of the Air Force Chief of Staff, General White; House, *Missile Development and Space Sciences: Hearings before the Committee on Science and Astronautics*, 86th Congress, 1st sess., February and March 1959, 76-77. The criticism has endured ever since.

importance to the nation as air, land, and sea power.[58] The disparity between the Air Force's position and that of its political and joint leadership is striking. Unfortunately, as General Horner pointed out, there is no single leader of the military space effort. Consequently, space strategy evolves from a highly politicized, bureaucratic process. This is likely to change soon.

Two events occurred in January 2001, that will likely move American space strategy in the direction toward the control doctrine. The first occurred on 11 January with the release of the Space Commission Report.[59] The report identifies several known threats to American use of space systems, including Russian GPS jammers, and formally recommends that "the US must develop the means to both deter and to defend against hostile acts in and from space."[60] Furthermore, the report states, "The Commissioners believe the US Government should vigorously pursue the capabilities called for in the National Space Policy to ensure that the President will have the option to deploy weapons in space to deter threats to and, if necessary, defend against attacks on US interests."[61] This will include the ability to negate hostile satellite threats.[62] Although the Space Commission refers to this as part of a larger deterrence policy, nonetheless it has a space control doctrine at its core.

The second event occurred on 20 January with the inauguration of President George W. Bush. President Bush is likely to abandon formally the sanctuary doctrine in favor of a space control doctrine with some elements of a high-ground doctrine. In the opening

---

[58] National Space Policy, The White House, 19 September 1996, 1-12. Department of Defense Space Policy, 9 July 1999, 2. Department of Defense, *Annual Report to the President and to the Congress, 2000* (Washington, D.C.: Government Printing Office), 84.
[59] The report is formally titled "The Report of the Commission to Assess United States National Security Space Management and Organization."
[60] Ibid., 100.
[61] Ibid., xii.

weeks of his administration, he immediately began strongly advocating the development and fielding of a missile defense system, which some members of the media and Congress believe implies basing weapons in space.[63] Also, his appointment to the position of Secretary of Defense was Donald Rumsfeld, the former chairman of the Space Commission. This increases the likelihood that the DoD will implement the Space Commission's recommendations. Right away, Secretary Rumsfeld indicated his desire for the US to renegotiate with the Russians or withdraw from the Anti-Ballistic Missile Treaty, which he called "ancient history," to facilitate the missile defense program.[64] *It now appears that given the findings of the Space Commission and a more realist-centered Bush administration, that the US will likely pursue the space control doctrine and possibly elements of the high-ground school as well.*

Alternative Strategies

There are alternatives to adopting a space control or high-ground doctrine that require consideration by policy makers. First, it may be most appropriate to regress back to the sanctuary doctrine by weaning military and commercial users from the space support to which they have grown accustomed. This will close the vulnerabilities that space dependence breeds and prevent the storm that will invariably occur should the US trigger security dilemmas while pursuing a more aggressive space policy. As appealing as this might sounds, it rests on the assumption that the US will have uncontested space superiority for at least the next ten to fifteen years and that the Services will be willing to

---

[62] Ibid., 29.
[63] James Dao, "Rumsfeld Plan Skirts Call for Stationing Arms in Space," *New York Times*, 9 May 2001, on-line, Internet 15 May 2001, available from http://ca.dtic.mil/cgi-bin/ebird?doc_url=May2001/e20010509rumsfeld.htm.

fall back to terrestrial-based technologies. Second, the US could covertly adopt a space control doctrine and secretly field the necessary systems while preserving the appearance of abiding by the sanctuary or survivability doctrines. If secrecy endures, this would prevent security dilemmas, however it risks far greater diplomatic repercussions if other nations discover the truth. It also would hamper the on-going force integration efforts. Secrecy would likely prevent space control systems from integrating effectively into real-world operations in much the same way secrecy prevented integration in Desert Storm and earlier operations. Third, the US could improve the durability of its satellites to allow them to withstand attacks as a continuation of the survivability doctrine. This is by no means a perfect solution, but it does add a degree of assurance that friendly space support will be present when needed. Fourth, the US could add a space denial strategy to its survivability doctrine. Space denial would involve traditional air, land, sea, special operations, and information attacks in a counter-force effort to deny an enemy access to space support by destroying its space launch and ground control infrastructure. It may also include striking whatever ground-based anti-satellite weapons they might possess. It is analogous to a blockade. This strategy is entirely possible using current force structures, but suffers from the same weakness as blockades at sea; they can be broken down, circumvented, or the adversary can receive space support from third parties.[65] Finally, the US could pursue a combination of the aforementioned ideas.

---

[64] Steven Lee Myers, "Bush's Choice for Defense Sees Immediate Bid To Raise Spending," *The New York Times*, 12 January 2001.
[65] Sheldon, 4.

39

Conclusions

During the Cold War, the US pursued the sanctuary doctrine, securing overflight rights of spysats which provided an element of stability in midst of an anarchic international system. However, the end of the Cold War and advancements in technology allowed the US to quickly migrate toward the survivability doctrine—focusing on space-derived force enhancement capabilities. However, the increased national reliance on space systems opened new vulnerabilities to potential adversaries who are fielding credible space threats. Accordingly, General Eberhart called for a move towards a space control doctrine, however this was contrary to the policies of the Clinton administration. His call has likely fallen on more fertile soil in the new Bush administration, which appears to embrace a new role for spacepower in America. With the former chairman of the Space Commission sitting as the Secretary of Defense, and several Congressionally sponsored studies citing America's growing reliance upon its increasingly vulnerable space assets, changes in space policy and doctrine are more likely now than at any other time in recent years.

The turn of the 21st century finds America at a crossroads in space. What course America ultimately pursues will likely have far reaching effects that will resonate for many years. In these defining moments, it is essential for the people of a spacefaring nation to contemplate some basic propositions regarding spacepower.

Chapter 3

Ten Propositions Regarding Spacepower

If you develop rules, never have more than ten.

—Sec Def Donald Rumsfeld
19th Rule of "Rumsfeld's Rules"

The objective of these propositions is to help train the judgments of political and military practitioners regardless of their organizational affiliations. Strategists and operational planners will find these propositions particularly helpful in developing their understanding of spacepower. Largely, these propositions are ten things to keep in mind when building spacepower for the nation, or when employing spacepower in a strategic, or operational sense to secure national or military objectives. The particulars of tactical employment are left for others to ponder.

Nearly fifty years of spacefaring experience has delivered many lessons regarding spacepower. Unfortunately, these lessons frequently disappear from view because of organizational interests. This is not to say that evil people are plotting to keep spacepower from rising to its full potential out of organizational self-interest. The fact remains that access to space is relatively new. Until Desert Storm, practically no one was thinking about spacepower as an element of military and national power. This has changed, but space operations remain very expensive, complex, and somewhat of a mystery to all but a relatively small group of operators, some generals, and even fewer political leaders. The organizations in which spacepower is currently vested have their own long-standing priorities and bureaucratic inertia that changes slowly over time.

Assessing these issues was the mission of the Space Commission, whose findings and recommendations appear in Appendix E of this study. These propositions simply provide a coherent way of thinking about spacepower for the nation.

Here, then, are the ten propositions regarding spacepower:

1. Space is a Distinct Operational Medium
2. The Essence of Spacepower is Global Access and Global Presence
3. Spacepower is Composed of a State's Total Space Activity
4. Spacepower Must Be Centrally Controlled By A Space Professional
5. Spacepower is a Coercive Force
6. Commercial Space Assets Make All Actors Space Powers
7. Spacepower Assets Form a National Center of Gravity
8. Space Control is Not Optional
9. Space professionals Require Career-Long Specialization
10. Weaponizing Space is Inevitable

#1 Space is a Distinct Operational Medium

To land, sea and air may now be added infinite empty space as an area of future intercontinental traffic, thereby acquiring political importance. This third day of October 1942 is the first of a new era of transportation— that of space travel.

—Major General Walter Dornberger, V-2 Project Director
In remarks to his staff upon the first successful flight of the V-2

We airmen who fought to assure that the United States has the capability to control the air are determined that the United States must win the capability to control space.... There is no division...between air and space. Air and space are an indivisible field of operations.

—General Thomas D. White, USAF Chief of Staff, 1957

Space is a medium like the land, sea, and air within which military activities shall be conducted to achieve U.S. national security objectives.

—DoD Space Policy, 9 July 1999

Space is a medium like the land, sea, and air where military activities are conducted.

—William S. Cohen, Secretary of Defense

Space is not simply a place from which information is acquired and transmitted or through which objects pass. It is a medium much the same as air, land or sea.

—Space Commission Report (Rumsfeld II), 11 January 2001

A significant milestone in human achievement occurred on 2 November 2000. On that date, astronauts took up permanent residence aboard "Alpha," the International Space Station. In that sense, space is much more like the land upon which people live than the oceans or the air through which people traverse. In another sense, Alpha is like a ship at sea that never returns to port, but swaps out its crew at regular intervals. The astronauts aboard Alpha will serve four-month tours of duty, similar to the duration of routine overseas deployments served by American airmen.[66] The bottom line is that "space is a place, not a program."[67] It is quite literally a place that is out of this world.

The proposition that space is a distinct operational medium may seem intuitively obvious, but the assertion that space is somehow operationally different from the air is anathema to many American airmen who believe the "military arts" dictate that air and space form a "seamless medium unconstrained by arbitrary divisions of the vertical dimension."[68] This is the central argument of the aerospace integration school of thought.[69] However, the divisions between air and space are not arbitrary. In reality, there are distinguishable physical, operational, and political divisions between air and

---

[66] Todd Halvorson, "Destination: International Space Station," *Space Views*, 30 October 2000, on-line, Internet 4 March 2001, available from http://www.spaceviews.com/2000/10/30a.html.

[67] Herbert F. York, *Making Weapons, Talking Peace: A Physicist's Odyssey from Hiroshima to Geneva* (New York: Basic Books, 1987), 160.

[68] Maj Gen John L. Barry and Col Darrell L. Herriges, "Aerospace Integration, Not Separation," *Aerospace Power Journal* 14 no. 2 (Summer 2000): 42.

[69] For a synopsis of the aerospace integration perspective, see Barry and Herriges, 42-47.

space that neither rhetoric nor technology can resolve. Failure to appreciate the differences between these two distinct media not only hobbles both air and space power from developing their full potential, but more importantly, may lead to serious political and military consequences if spacepower is misapplied as if it were merely an extension of airpower.

Air Force doctrine recognizes the physical differences between the air and space environments, but it denies any separation between them.[70] To the contrary, a separation becomes apparent when we consider the limitations that physics imposes on practical applications in the two media. In doing so, we realize that the upper boundary of the air medium is 28 miles, the highest altitude attainable by an air-breathing aircraft.[71] Conversely, the lower boundary of space is approximately 93 miles, the lowest possible perigee for a satellite. These functional limitations make it obvious that air and space do not meet. There is a region nearly 65 miles wide that divides air from space. This is the *transverse region*, wherein neither aerodynamic flight nor orbital rotation is possible. Despite the Air Force's insistence of an "aerospace continuum," *the transverse region forms an invisible barrier 65 miles wide that divides air operations from space operations.*[72]

Operating inside the transverse region is not practical because the energy expenditures required to maneuver or loiter there are too great—by orders of magnitude. Vehicles can exploit neither Bernoulli's aerodynamic principles nor Kepler's

---

[70] Air Force Doctrine Document (AFDD) 2-2, *Space Operations*, 23 August 1998, 1.
[71] The X-15 achieved an altitude of 67 miles on 13 September, 1968, but this vehicle was rocket powered and glided ballistically to this altitude after expending all of its rocket fuel at a much lower altitude.
[72] The term "aerospace continuum" is used repeatedly by Hon F. Whitten Peters and Gen Michael E. Ryan, in *America's Air Force: Global Vigilance, Reach, and Power* (Washington D.C.: Department of the Air Force, June 2000), and subsequently throughout Air Force publications.

astrodynamic principles to conserve energy. Consequently, the transverse region lives up to its name as a boundary across which vehicles transit, but do little else. The great cost of space operations hinges on spacelift vehicles—large rockets—that generate the huge amounts of energy required to lift payloads through the transverse region and accelerate them to orbital speed and altitude.[73]

Claims that technology will eventually overcome the limitations imposed by the transverse region, thereby enabling vehicles to operate with agility in both media, rest on optimistic speculation. To date, the wings mounted on space vehicles, such as the Space Shuttle, Space Maneuvering Vehicle, and other experimental craft currently in development, are merely recovery systems analogous to parachutes. One wonders how much extra time, money, and payload tradeoff went into designing winged recovery systems and their ground and airspace support infrastructures, when a parachute system would weigh less and provide a similar effect—safe return to Earth.

One of the primary reasons the Air Force resists recognizing space as something separate from the air is to avoid creating "arbitrary lines on a map" that create command and control problems similar to the Army/Air Force controversy surrounding the fire support coordination line (FSCL).[74] The fear is that establishing space as a geographic area of responsibility and as a distinct theater of operations will somehow hinder the freedom of air and space assets to maneuver and employ weapons because of command and control issues.[75] Given the wide separation between air and space imposed by the

---

[73] To give an idea of how much energy is required to cross the transverse region, scientists calculate that ninety percent of the total energy required accomplishing a mission to Mars is spent lifting the payload into a low Earth orbit.

[74] Maj Gen Lance Smith, Air Force Doctrine Center, Maxwell Air Force AFB, Ala., interview with author, 9 March 2001.

[75] Ibid.

transverse region, this seems unlikely to pose a major problem. In fact, to a large degree, the Air Force has already lost this argument because separate coordination with space forces is already required. "Any Department of Defense agency wanting to fire a laser [or any other weapon] above the horizon [into space] must first get permission from US Space Command's Laser Clearing House, which uses the center's information on satellite vulnerabilities to help determine if there are any satellites in the [weapon's] path that should be avoided."[76] Whether or not the Department of Defense ever recognizes space as an AOR or theater, it appears a FSCL is already in place.

Lieutenant General David Vesely, Vice Chairman of the Joint Chiefs of Staff, expressed another core belief of the aerospace integration school when he stated, "Whatever differences there are between air and space are *not* important to the theater commander (strategic level) or the warfighter (tactical level). What is important is the *effect* on the battlefield. Whether it's weapons, communications, or information, the warriors out there don't care where it came from as long as it has the desired impact on the battlefield."[77] Other generals insist, "At the operational level of warfare, air and space are absolutely identical."[78] Are they claiming that air and space are essentially the same at all levels of warfare? There seems to be some confusion in the minds of aerospace integrationists about exactly where the art and science of air and space power converge within the seamless medium they have constructed.

---

[76] Rich Garcia, "Lab Evaluates Satellites," United States Air Force News Release, 6 July 2000, n.p.; on-line, Internet 10 March 2001, available from http://www.sor.plk.af.mil/pa/releases/2000/00-50.html.
[77] Quoted in Tech Sgt Ginger Schreitmueller, "Assistant Chief of Staff Talks About AEF Capabilites, Resources," *Air Force News*, 31 July 1998. (emphasis added) A very close paraphrase of this quote is attributed to General John Jumper in white paper by the Hon F. Whitten Peters and Gen Michael E. Ryan, *The Aerospace Force: Defending America in the 21st Century* (Department of the Air Force, Washington D.C., 2000), 3. This suggests collaboration among the senior leaders of the Air Force to "stay on message" with the Chief of Staff's policy, or at least collusion between speechwriters.
[78] Smith interview.

While it is certainly true that tactical level warfighters do not care where *support* comes from, they are particular about the type of support they receive. Some aircraft and spacecraft perform similar missions, i.e. reconnaissance, surveillance, communications, etc, but the type of support each can deliver is different. This is because aircraft and spacecraft operate in different media with vastly different ranges and speeds relative to targets, and are constrained by different physical laws of motion. This means the differences between airpower and spacepower *are* important to theater commanders, component commanders, and operational planning staffs—*at least they ought to be!* These people are involved with creating the plans that connect the ends with the means in warfare. If the means are different, so must be the plan.

The dramatic difference between planning the employment of aircraft on one hand, and spacecraft on the other, highlights the fact that airpower and spacepower presents war planners with different means to accomplish the ends they seek. At the tactical level of warfare, airmen and space professionals practice entirely different sets of warfighting skills—flying an aircraft in theater is different from remotely commanding a global network of satellites from the continental US. At the operational level of warfare, Air Operations Centers around the globe focus almost exclusively on the air situation in their respective theaters. Conversely, the one and only Air Force Space Operations Center at Vandenberg Air Force Base, California, has to maintain a global perspective and tasks its assets to support users worldwide.[79] At the strategic level of warfare, combined/joint force commanders obtain their space support via a supported/supporting relationship

---

[79] Actually, the terms "Air Operations Center" and "Space Operations Center" are passé. Both are now referred to as "Aerospace Operations Centers." This is arguably more of a superficial change than a change of substance.

from USCINCSPACE. At the grand strategic level of national policy, The National Command Authority (NCA) frequently tasks space assets independently of other forces in the Department of Defense (DoD) to achieve political ends.[80]

The physical remoteness and the laws of orbital motion in space create several defining operational characteristics of spacepower that separate it from other forms of power. In his book *Modern Strategy*, Colin S. Gray points out several of the inherent advantages associated with operating above the transverse region in the space environment:

> First, space is but the latest variant of the 'high ground' that doctrine often advises military commanders to seize and hold. As with forces on all kinds of high ground, space systems look down on friend and foe and are relatively difficult to reach and grasp. To attack uphill had never been easy; to attack up the Earth's gravity well would continue that military condition. Second, the high ground of space is both global and of all but infinite military depth. The country or coalition which can operate at will in space is able thereby to operate from the highest vantage points. And finally, space power, obedient to Keplarian laws of orbital motion, translates as satellites that can be available globally as either a regularly repeating or a constant overhead presence."[81]

Gray also points out some of the operational limitations imposed by the space environment:

> Spacepower has several limitations. The high cost of transportation into orbit (i.e. launch costs) limits the pace of advance of military, scientific, and commercial space systems... Next, the laws of orbital motion that govern celestial bodies are a permanent constraint upon the flexibility with which spacepower can be employed; those laws can be overridden to a degree, but only with a virtual attrition in payload imposed by the fuel necessary to achieve some anti-Keplarian agility... Predictability 'on orbital station', or predictable orbital passage, is both a blessing and a vulnerability. The orbital task calculated by us to provide the necessary

---

[80] The NCA frequently requests specific space derived intelligence products from the National Reconnaissance Office. In addition, the NCA also tasks DoD space assets directly to achieve political purpose, as was the case with the decision to terminate selective availability of the Global Positioning System in 2000.
[81] Colin S. Gray, *Modern Strategy* (New York: Oxford University Press, 1999), 260-261.

terrestrial support is also calculable by the anti-satellite weapon systems of the foe. [Finally,] it is in the nature of spacepower to be distant from terrestrial events. Although it is the distance overhead that is militarily beneficial, still distance from Earth is an important limitation.[82]

The issues that Gray raises about operating in space are similar to those faced by practitioners in the other media. Operators in all media are concerned about the high ground, presence, cost, maneuver, defensibility, and range. Gray argues that these similarities exist because "spacepower…is not governed by a distinctive strategic logic," rather it follows the same strategic logic that applies to any independent medium of operations.[83] A listing of Gray's other insightful observations regarding space as it relates to its nature as an operational medium can be found in appendixes B and C of this study.

The advantages and limitations of spacepower that Gray identifies are no surprise. Space professionals understood these enduring characteristics of space operations at the outset of the space age. Many authors have addressed these characteristics as a way to appreciate the differences between air and space operations. In 1995 Lieutenant Colonel Michael R. Mantz concluded in his theory of space combat power:

> Air and space are operationally different. Aircraft have maximum maneuverability, while spacecraft have greater altitude and speed, but can't maneuver [with even a fraction of an aircraft's agility and flexibility]. The principles of war of mass and maneuver certainly do not apply in the same way. Aircraft can mass repeatedly through maneuver over a target, while spacecraft can mass for short periods after great effort, but will disperse almost immediately with a repeat manning unlikely. Aircraft operations are "on demand," while spacecraft operations are "as scheduled" or "when available."[84]

---

[82] Ibid., 263-264.
[83] Ibid., 260.
[84] Lt Col Michael R. Mantz, *The New Sword: A Theory of Space Combat Power* (Maxwell AFB, Ala.: Air University Press, May 1995), 79-80.

Operations in the space medium therefore require their own concept of operations—integrated with all other forms of military power—but fully accommodating the unique physical and operational attributes of spacepower.

The politics of space are perhaps the most important distinction between space and the other media. Numerous space-specific treaties, agreements, and arrangements of customary law politically separate space from the terrestrial media. If air and space are truly a seamlessly integrated medium of operations, then why do spacecraft have unrestricted overflight rights whereas aircraft do not? Why is not an aircraft carrying bombs aloft considered the weaponization of space? Why did the United Nations pass a non-binding resolution in November of 1999 calling upon governments to "contribute actively to the prevention of an arms race in outer space [no mention of air] and to refrain from actions contrary to that objective?"[85] The answers to these questions are rooted in the fact that the international community understands space as a separate and distinct operational medium. Different political rules apply to space than air. Among the space-specific treaties to which the US has agreed:

- The 1963 Limited Test Ban Treaty prohibiting "any nuclear weapon test, explosion, or any other nuclear explosion in outer space."

- The 1967 Outer Space Treaty proscribes placing weapons of mass destruction in space or on the moon or other celestial bodies for any military purposes.

- The 1972 Anti-Ballistic Missile (ABM) Treaty prohibits the development, testing, or deployment of space-based components of an anti-ballistic missile system.

---

[85] United Nations, "Assembly Urges Enrichment of United Nations 'Internet' Presence, Full Use of Emerging Information Technologies," United Nations Press Release GA/9677, 6 December 1999.

- A number of arms control treaties are intended to prohibit the US and Russia from interfering with the other's use of satellites for monitoring treaty compliance.

- The 1980 Environmental Modification Convention prohibits all hostile actions that might cause long-lasting, severe, or widespread environmental effects in space.[86]

The political implications of military spacepower are quite apparent—a powerful international lobby wants to preserve space as a peaceful sanctuary. Many fear that conducting offensive operations in, from, or through space has the potential to destabilize the global power structure thereby causing unpredictable security dilemmas. Several nations voice this concern regularly through the UN Secretariat of the Conference on Disarmament, regarding proposals to place weapons in orbit.[87]

If Clausewitz is correct that war is an extension of politics, and space has a unique set of political considerations associated with its use, then military planners at all levels of warfare must pay very careful attention to employ spacepower consistent with the political aims of the war.[88] In most cases, this will be in a manner quite different from airpower. Operations in the space medium will likely require their own rules of engagement apart from those guiding other forms of military power because of the unique set of political constraints regarding space.

The physical, operational, and political nature of space set it apart from the terrestrial media. These factors are especially important to military planners at all levels of warfare because spacepower has several unique planning considerations that must be taken into

---

[86] *Report of the Commission to Assess United States National Security Space Management and Organization* (Washington D.C., January 11, 2001) 37.
[87] Pericles Gasparini Alves, *Prevention of an Arms Race in Outer Space: A Guide to the Discussions in the Conference on Disarmament*, UNDIR/91/79, Annex A (New York: United Nations Institute for Disarmament Research, 1991), 2.

account when connecting the ends with the means to achieve the desired effects (objectives) in the battlespace. This makes space a distinct operational medium.

#2 The Essence of Spacepower is Global Access and Global Presence

> *This is Friendship 7...Zero 'G' and I feel fine...Oh, that view is tremendous!*
>
> —Mercury Astronaut John Glenn
> 20 February 1962

Like most Americans, President Eisenhower could never forget Pearl Harbor. "As president, his scientific advisor, James Killian, remarked that Eisenhower remained 'haunted'...throughout his presidency' by the threat of surprise nuclear attack on the United States."[89] Indeed, a survey conducted in the mid-1950s indicated that more than half of all Americans believed they were more likely to die in a nuclear attack than from old age.[90] Feeding this sense of paranoia was the closed nature of the Soviet Union, which as Eisenhower noted, gave them an advantage in planning a secret attack when compared to the open American society.[91] The problem facing the administration was how to gain access to the Soviet Union in order to assess their military capabilities and intentions.

Ultimately, the Eisenhower administration tried four methods of peeking behind the iron curtain. First, at a July 1955 Geneva Four Power Summit Conference, the President

---

[88] More precisely, Clausewitz stated that "war is simply a continuation of political intercourse, with the addition of other means." Carl von Clausewitz, *On War,* ed. and trans. by Michael Howard, and Peter Paret (Princeton, N.J.: Princeton University Press, 1984), 605.

[89] James R. Killian Jr., *Sputnik, Scientists, and Eisenhower: A Memoir of the First Assistant to the President for Science and Technology* (Cambridge: The MIT Press, 1977), 68. Quoted in David N. Spires, *Beyond Horizons* (Wasington D.C.: Government Printing Office, 1997), 30.

[90] Curtis Peebles, *High Frontier: The US Air Force and the Military Space Program* (Washington, D.C.: Government Printing Office, 1997), 3.

[91] Ibid., 4.

proposed an initiative called "Open Skies," wherein reconnaissance aircraft of the US and Soviet Union would freely overfly each other's country.[92] The Soviets rejected this initiative. Second, in January 1956, Eisenhower authorized the release of reconnaissance balloons to overfly the Soviet Union under the guise that they were merely weather balloons. This practice ended a few days later amidst serious Soviet protests and very few useful photographs recovered.[93] In 1958, America released balloons again, but this effort quickly met the same results. Third, in July 1956 the President authorized overflights by the newly operational U-2 reconnaissance aircraft. This was a great success until May of 1960, when the Soviets shot down Gary Powers in a U-2 over their homeland. In the political aftermath, the US terminated all future overflights of the Soviet Union. Shortly thereafter, Soviet space law theorist, Georgi Zhukov, warned in October 1960 that since the USSR had proved it could shoot down American spy planes, that "the United States would rush development of a new method via satellites in space."[94] How right he was. Satellite reconnaissance was the forth method the Eisenhower administration pursued in order to gain intelligence about the Soviet Union, but the effort to build a spysat had begun some years earlier.

Since 1946, scientists seriously contemplated reconnaissance satellites. In that year, the RAND Corporation published their first study titled, "Preliminary Design of an Experimental World-Circling Spaceship." "A 'satellite offers an observation aircraft which cannot be brought down by an enemy…,' the report observed. Other military roles

---

[92] Ibid., 4-5.

[93] Col Delbert R. Terrill, Jr., *The Air Force Role in Developing International Outer Space Law* (Maxwell AFB, Ala.: Air University Press, May 1999), 5.

[94] Georgi P. Zhukov, "Space Espionage Plans and International Law," International Affairs (October 1960). Quoted in Walter A. McDougall, …The Heavens and the Earth (Baltimore: Johns Hopkins University Press, 1985), 259.

included the 'spotting of points of impact of bombs launched by us, and the observation of weather conditions over enemy territory.'[95] Over the next decade, politicians and scientists alike would come to appreciate the idea of satellites as a means to access denied geographic regions, such as the Soviet Union. This prompted the Eisenhower administration to seek a way to establish freedom of international overflight rights for satellites (an logical extension of their "Open Skies" policy). While the administration intended to establish the precedence of overflight rights by launching satellites as part of the International Geophysical Year, planned for late 1957-1958, the Soviets actually did most of the work when they shocked the world with the launching of Sputnik on 4 October 1957. From that moment on, the freedom of passage in space became customary law, and later international law.[96] In August of 1960, just three months after the U-2 overflights ended, satellites began returning images from inside the Soviet Union. This information allowed political and military leaders to shape their diplomatic and military efforts more effectively to address the real threat and not some perceived threat based on guesswork. Space was no longer merely a science project, but an instrument of policy in the Clausewitzian sense. True spacepower had arrived.

The driving reason for America's initial voyage into space was to exploit the unique element of global access, an inherent attribute of most low Earth orbiting satellites.[97] Jim Oberg, author of *Space Power Theory* noted:

---

[95] Cited in Peebles, 1.

[96] The Outer Space Treaty of 1967, signed by 90 other countries, including the US and USSR, established the idea that space is for peaceful purposes only and that "passage through space…must be free from interference."

[97] Reconnaissance satellites are typically placed in a sunsynchronous orbit, meaning the satellite is placed in a retrograde orbit relative to the Earth's rotation and the satellite's South-to-North pass has a constant angle to the sun. In this type of orbit satellites typically complete one revolution every 70 to 120 minutes and repeat the same ground track every few days.

The primary attribute of [Earth focused] space systems lies in their extensive view of the Earth. Ability to service large areas from a distance of less than a thousand kilometers for most low-Earth systems is the key ingredient for stationing the vast majority of systems in space. It is this extended area—virtually global in nature—that not only represents spacepower's most valuable asset, but also sets it apart from all other forms of power. While all other forms of power are effectively regional, spacepower allows worldwide access in time spans measured in minutes as opposed to hours and days.[98]

Closely related to the attribute of global access is the attribute of global presence. By placing several satellites of a particular type into orbit at certain altitudes and distribution, we create a "constellation" of satellites. The more satellites added to the constellation, the more coverage of the Earth's surface, but another factor is the constellation's altitude above the Earth's surface. For example, the Iridium Satellite Company operates 66 satellites in a low Earth orbit (485 miles) that provide telecommunications worldwide.[99] Still higher, at a medium Earth orbit (11,000 miles), the US Air Force's Global Positioning System employs 24 satellites to maintain global coverage while ensuring all points on Earth are in view to at least four satellites. Finally, at a much higher geostationary orbit (22,300 miles), only three satellites are required to provide missile warning surveillance or communications over most of the globe. At this altitude, a single satellite views nearly 40 percent of the Earth's surface.[100] However, for

---

[98] James E. Oberg, *Space Power Theory* (Washington, D.C.: Government Printing Office, March 1999), 124.

[99] "How Iridium Works" Iridium Satellite LLC, n.p.; on-line, Internet 8 April 2001, available from http://www.iridium.com/

[100] One might assume that a satellite in a geostationary orbit might see half of the Earth's surface, like an astronaut viewing a full Earth rising from the surface of the moon. This would mean only two satellites would be required to provide global coverage. In reality, geostationary satellites have a line of site limitation around the edges of the Earth sphere that accounts for the 10 percent reduced viewing area. Even when three satellites are employed in geostationary orbits, there is still a loss of coverage at the polar regions, but this can be compensated for by adding satellites in a Molnaya (highly elliptical) orbit that provides extended dwell time over those regions.

the sake of redundancy and overlapping coverage areas, more than three satellites are typically used.

When selecting an orbital altitude for a particular satellite or constellation of satellites, there is a tradeoff between sensor range to the surface of the Earth and the total number of satellites needed to provide global coverage. Traditionally, reconnaissance and other Earth sensing satellites are placed in low Earth orbits in order to take advantage of the relatively close range to the surface. In order to provide global presence with these types of satellites, constellations of several hundred satellites would be required—at an enormous expense. As a result, nations have not fielded reconnaissance satellites in sufficient numbers to yield global omnipresence; instead, they find it adequate to accept frequent revisits of these satellites to all areas of interest.

For missions that require global coverage, i.e. navigation, communications, and missile warning, economic factors typically drive the purchase of a few satellites placed in higher orbits to achieve global coverage. There is another reason to place satellites on the geostationary belt, at 22,300 miles above the equator. Only on the geostationary belt can satellites remain in one spot relative to the surface of the Earth. This is particularly useful for large communications systems because users on the surface of the Earth can keep their antenna fixed in one position, instead of having to track the movements of particular satellites while perpetually moving their antenna accordingly. Customers of telecommunications systems that employ several satellites in low Earth orbit, such as Iridium, get around the problem of tracking satellites and adjusting antennas by using very low power handsets that use omni-directional antennas to reach any one of the satellites in the constellation.

The essence of spacepower is global access and global presence. These attributes are often the answer to the question "why do we go to space?" They are the reasons for performing most military, civil, and commercial missions that are performed in space. This point was widely recognized only a few years ago. In the current Air Force Doctrine Document 2-2, Space Operations, dated 23 August 1998, a section titled "Global Coverage" provides the following explanation:

> Space-based systems in appropriate orbital deployments provide worldwide coverage and frequent access to specific Earth locations, including those denied to terrestrial-based forces, on a recurring basis. Unconstrained by political boundaries, satellites deployed in specific orbits and in sufficient numbers maintain a continuous presence over enemy [and friendly] territory…Space systems provide an instantaneous presence not available from terrestrial-based forces, permitting the United States to leverage information to influence, deter, or compel an adversary or affect a situation. The use of multiple space platforms allows warfighters to exploit the various sensors, resulting in a synergistic battlespace perspective that reduces the fog of war. Although space systems provide global coverage, some can be focused to provide information on specific areas of interest, which can improve situational awareness and planning tempo and can enable information dominance for all friendly military forces. By exploiting comprehensive space capabilities, space forces can focus on and provide detailed services for a specific geographic area and support regional planning requirements. The attribute of global coverage offers significant advantage to Air Force battle management. Properly positioned in sufficient numbers, space-based systems could provide a global capability for much of the information currently provided by airborne platforms such as the joint surveillance, target attack radar system (JSTARS) and the Airborne Warning and Control System (AWACS).[101]

There is a great risk that the Air Force is losing sight of the essence of spacepower in its fervor to advance aerospace integration. A recent draft of space doctrine that is slated to supersede the current edition makes no mention of the unique global access and global presence of spacepower that set it apart from all other forms of power.[102] Doing so

---

[101] Air Force Doctrine Document (AFDD) 2-2, *Space Operations*, 23 August 1998, 15-16.
[102] AFDD 2-2 (Draft), Space Operations, (September 2000).

would undercut the aerospace argument that air and space are a seamless operational continuum.

In reality, the different attributes of airpower and spacepower mean they do different things, and whatever they do that is similar, is done differently. For example, aircraft do not overfly denied areas. Spacecraft do. Airpower does not provide a global presence. Spacepower does. Airpower is much closer to the fight than spacepower, but several capabilities provided by spacepower are always present during war and peace, whether or not terrestrial forces are present.

When an airman speaks of the global nature of airpower, he means something very different than when a space professional speaks of the global nature of spacepower. An airman means airpower has global *reach*. Global reach is the ability of an aircraft to takeoff from point-A and travel to point-B anywhere on the globe to create an effect (read as achieve an objective). In contrast, a space professional means some spacepower assets provide nearly ubiquitous presence around the entire globe with some capabilities, while other spacepower assets provide frequent revisits of all areas of the globe with other capabilities. Spacepower provides its effects across the globe all the time. This is because the essence of spacepower is global access and global presence.

#3 Spacepower is Composed of a State's Total Space Activity

Air power is not composed alone of the war-making components of aviation. It is the total aviation activity civilian and military, commercial and private, potential as well as existing.

—Gen Henry H. "Hap" Arnold

Space power is not composed alone of the war-making components of space. It is the total space activity; civil, commercial, defense, and intelligence, potential as well as existing.

—Coyote's Space Corollary to Arnold's Air Power

Going to space is hard. If it was easy mankind would likely have gone there long before Sputnik became the first manmade satellite in 1957.[103] Taking the initial steps into space were only possible because the competing superpowers were reacting to survival-level security dilemmas during the Cold War—and were willing to support huge government programs to get there. This makes space unique because government interest in the sea and air grew after civilian curiosity and private enterprise opened these terrestrial media. With more than four decades of spacefaring experience behind us, going to space is still hard and expensive, even though we have answered most of fundamental questions and resolved many of the technical problems.

In *Space Power Theory*, author Jim Oberg identifies several attributes of a spacefaring nation. He claims that "Several basic traits are shared by most spacefaring nations: geographical size and location, national wealth, an extensive and well-educated population, existing national power, a popular appetite for technology, and political

---

[103] Arguably, the Germans took the initial steps into space with their V-2 rocket program, which accomplished its first successful flight into sub-orbital space on 3 October 1942.

59

will."[104] He notes it is difficult to tell which of these attributes is the most important, but claims "spacepower can be conceived as a combination of all the quantitative factors multiplied by the qualitative factor of will."[105] It is immediately obvious that astate must have a great deal of intellectual, industrial, and capital slack in order to build a successful space program. In addition, the internal political environment must tolerate such expenditures. A government that must spend every available penny on social programs is not likely to develop or sustain a space program.

In short, many things must be going right inside a state for it to afford the investment in a start-to-finish space program. The Russian space program, possibly the leader in the late 1950s and early 1960s, faltered in the wake of the break-up of the Soviet Union in the early 1990s. Their social needs outweighed the urgency for the continued investment required to reconstitute their satellite constellations and meet their fiduciary commitments to the International Space Station. Nearly ten years after the break-up, it appears they have worked-out many of their earlier problems and may be ready for resurgence in space. Other nations, most notably China and perhaps India, have finally developed a national infrastructure wherein they can realistically set the goal of conducting manned space missions.[106]

Spacepower in the US migrated quickly from its defense related origin and evolved into four distinct sectors of space activity: civil, commercial, defense (also referred to as military), and intelligence. When considered together, activities in these four sectors

---

[104] James E. Oberg, *Space Power Theory* (Washington, D.C.: Government Printing Office, March 1999), 131.

[105] Ibid., 27.

[106] "The People's Republic of China recently announced its intention to become the third nation to place human beings into orbits and return them safely to Earth." Cited in *Report of the Commission to Assess United States National Security Space Management and Organization* (Washington, D.C.: Government Printing Office, 11 January 2001), 10. (henceforth Space Commission Report)

describe the *national spacepower* of the US, and become a useful template to describe the

national spacepower of other states. Peter Hays, et al., described these four sectors in

Spacepower for a New Millennium like this:

> The US Government conducts *civil* space activities in order to explore the universe and advance human knowledge. The National Aeronautics and Space Administration (NASA) performs these missions, and they are funded by the government. Examples include human spaceflight missions conducted under the Apollo, Skylab, and International Space Station programs; robotic exploration performed by programs such as Viking, Voyager, Galileo, and Mars Pathfinder; and scientific missions of the Earth Observation System or Landsat programs. These types of civil space missions are probably the first space activities to come to mind for most people, but they are only the tip of the iceberg of all space activities.[107]
>
> *Commercial* space activities are performed by the private sector to make money. Communications satellites and telecommunications services form the oldest and most profitable segment of the commercial space sector. Other commercial space services that are or soon may become profitable include: navigation and positioning, launch, and remote sensing. Many commercial space sector activities are highly volatile economically and are governed by a complex set of international and domestic legal regimes such as those established by the 1967 Outer Space Treaty and the Commercial Space Act of 1998. These legal regimes are enforced by international and domestic regulatory agencies such as the International Telecommunications Union (ITU) and the Federal Aviation Administration.[108]
>
> The *intelligence* space sector consists of surveillance and reconnaissance missions designed to collect information for use by the US Government [and its allies]. Throughout the Cold War, space-based intelligence gathering activities, such as the Keyhole (KH) photo reconnaissance satellites, and the organizations involved in procuring and operating these systems, such as the National Reconnaissance Office (NRO), were highly classified or "black" due to the political sensitivities and cutting-edge technologies involved in this sector. Spy satellites are often given great credit for helping to stabilize the superpower relationship during the Cold War. They were the most important national technical means of verification (NTMV) for arms control agreements prior to the advent of on-site inspections (OSI) that were first allowed as part of the

---

[107] Peter L. Hays, et al., eds., *Spacepower for a New Millennium: Space and US National Security* (New York: McGraw-Hill, 2000), 2.
[108] Ibid.

Intermediate-Range Nuclear Forces (INF) Treaty of 1987 and, today, are critical for a wide range of increasingly complex missions beyond verification.[109]

The *military* space sector contains all the remaining national security missions. These are directed by the Department of Defense (DoD), commanded by USSPACECOM, operated primarily by Air Force Space Command (AFSPC), and provide data streams to enhance the effectiveness of US and allied forces worldwide. The vast majority of military sector systems and missions are now declassified, but some aspects remain black. Military sector missions cover a wide variety of actual and potential activities that range from supporting space operations to applying force to, in, and from space.[110]

Hays points out that there is substantial overlap between these four sectors.[111] This creates opportunities for the sectors to leverage-off one another to yield a synergistic effect. They often combine research dollars and readily adapt spin-off technologies from other sectors. Frequently they share facilities and resources, such as the Eastern and Western ranges and the Air Force Satellite Control Network (AFSCN). More importantly, they provide redundancy in several critical areas, such as satellite commanding, leasing commercial assets to cover requirements, debris avoidance warning, and space object tracking (space surveillance). This not only builds more robust national spacepower, but also creates a new dynamic that complicates space policymaking.

There will be challenges posed by the use of civil and commercial systems by the military sector. For example, is a commercial satellite system a valid target if it is providing information to an adversary? Is it legitimate to strike an adversary's satellite control network if it provides support services to international consortia? These questions and many others remain unanswered, but mixing commercial and civil assets

---

[109] Ibid.
[110] Ibid., 2-3.

together with military forces in the other media raises the same types of challenging questions.

The most important development in national spacepower has been the rise of the commercial sector. In the late 1990s, international commercial investment in space finally exceeded the combined spending of all governments in space. This marks a clear turning point. In the words of the Space Commission Report, "Unlike the earlier era, in which governments drove activity in space, in this new era certain space applications, such as communications, are being driven by the commercial sector. An international space industry has developed, with revenues exceeding $80 billion in 2000. Industry forecasts project revenues will more than triple in the next decade."[112] Space is now a place for business, capital investment, and more importantly, *profit*.

It seems likely that market forces will now dictate the cost of space programs, regardless of the sector. Although the military and intelligence sectors will undoubtedly have somewhat unique requirements in order to gain a technological edge on an adversary, it is likely that most of their systems or subsystems will be compatible with commercially available systems. The advantage of using a commercial off-the-shelf (COTS) systems is tremendous savings over having to develop an entirely new systems, as was the case only a few years ago.

The rise of the commercial space sector also brings space much closer to every human on the planet. We have no idea how deeply spacepower will touch the fabric of our lives. In the words of the Space Commission Report:

> Space-based technology is revolutionizing major aspects of commercial and social activity and will continue to do so as the capacity and

---

[111] Ibid., 3.
[112] Space Commission Report, 11.

capabilities of satellites increase through emerging technologies. Space enters homes, businesses, schools, hospitals and government offices through its applications for transportation, health, the environment, telecommunications, education, commerce, agriculture and energy. Space-based technologies and services permit people to communicate, companies to do business, civic groups to serve the public and scientists to conduct research. Much like highways and airways, water lines and electric grids, services supplied from space are already an important part of the U.S. and global infrastructures.[113]

However, unlike aviation pioneers who made the public more airminded by providing very exciting and tangible examples of airpower, such as air transportation, barnstorming (airshows), air races, etc, spacepower remains relatively hidden. This places a special challenge on all spacefaring states to keep the imaginations of their citizens sparked to the idea and potentials of spacepower, because only states of high science and energized industry can sustain a space program. This requires a permanent educational commitment to the hard sciences and the strategic arts that make it all possible. It is often quipped "No bucks, no Buck Rogers." The bucks begin with building a spaceminded nation because ultimately spacepower is composed of a state's total space activity.

#4 Spacepower Must Be Centrally Controlled by a Space Professional

[Space] warfare cannot be separated into little packets; it knows no boundaries on land and sea [or air] other than those imposed by the radius of action of the [space]craft; it is a unity and demands unity of command.

—Air Marshal Arthur Tedder
[space substituted for air]

Did we not learn anything in North Africa? When Operation Torch began in November 1942, airpower was neither independent nor centralized because senior army officers, both British and American, insisted on controlling their own airpower to provide

---

[113] Ibid., 10-12.

local protection and deal with *local* problems.[114] The results were disastrous. Many historians assert that the early failures of Torch, especially the Battle of Kasserine Pass, can be blamed on ground leaders who failed to listen to their airmen who advised them on controlling air resources.[115] The solution to the problem was giving airmen what they clamored for since the First World War; centralized control of airpower at the *theater* level by an airman with close liaison to Army headquarters.[116] Airmen went from taking orders to filling requests with the freedom to manage airpower in the most efficient manner that their expertise made possible.[117] Soon, with centralized control and close coordination with ground forces, the Allied air forces turned the tide on the Luftwaffe and hastened the defeat of Germany's Afrika Korps.[118]

Nearly sixty years later, the Air Force is making the same mistake with spacepower by promulgating in its basic doctrine that "air and *space* power must be controlled by an *airman* who maintains a broad strategic/*theater* perspective in prioritizing the use of limited air and *space* assets to attain the objectives of *all* US forces in any contingency across the range of operations."[119] The idea that an *airman* with a *theater* perspective should ever control space assets, which as space operations doctrine points out, are

---

[114] Vincent Orange, "Getting Together: Tedder, Coningham, and Americans in the Desert and Tunisia, 1940-1943," in Daniel R. Mortenson, ed., *Airpower and Ground Armies: Essays on the Evolution of Anglo-American Air Doctrine, 1940-1943* (Maxwell AFB, Ala.: Air University Press, 1998), 25.

[115] Daniel R. Mortensen, "The Legend of Laurence Kuter; Agent for Airpower Doctrine," Ibid., 93.

[116] David R. Mets, "A Glider in the Propwash of the Royal Air Force? Gen Carl A. Spaatz, The RAF, and the Foundations of American Tactical Air Doctrine," Ibid., 50-56.

[117] The air-land command relationship at the time of Kasserine flowed from FM 31-35, *Aviation in Support of Ground Forces*, adopted in April 1942. An outstanding summary of this failed command structure is found in Bernard C. Nalty's, *Winged Shield, Winged Sword: A History of the United States Air Force, Vol. 1* (Washington, D.C.: Air Force History and Museums Program, 1997), 271-272.

[118] Warren A. Trest, Air Force Roles and Mission: A History (Washington D.C.: Air Force History and Museums Program, 1998), 91.

[119] AFDD 1, Air Force Basic Doctrine, 1 September 1997, 23. (Emphasis added.)

properly "tasked and assigned from a *global* perspective," should send shivers up the backs of military leaders.[120] Think North Africa.

Here rises a conflict in the aerospace integration argument. Physical differences aside, air and space do *not* form a seamlessly integrated operational medium because airpower is best managed from a theater perspective while spacepower is best managed from a global perspective.[121] While airpower focuses on providing effects that support a combined or joint force commander, spacepower focuses on providing effects that support *all* joint warfighters and civil users, around the globe.[122]

Airpower is a theater-focused form of military power, which is why independent air components exist in every theater. So-called "global airpower" assets such as B-2 Stealth Bombers support only one theater at a time. Airpower has essentially achieved global reach with theater assets. Conversely, controlling space assets apart from the many pockets of airpower is necessary to allow space professionals to properly prioritize and optimize support to multiple users around the globe and deconflict necessary maintenance and sensor rest periods accordingly. These factors culminate in the need for space professionals to centrally control spacepower from a global perspective in order to achieve the greatest economy of force while supporting the highest priority needs of all users—especially the joint warfighters.

Space assets must not be broken into "penny packets" by allowing airmen (or any one else) in various theaters of operation to control whatever space forces are present or

---

[120] AFDD 2-2, 5. (Emphasis added.)
[121] Several recent documents assert the contrary point as part of the Air Force's aerospace integration initiative. See "The Aerospace Force: Defending America in the 21st Century" (Washington D.C.: The Department of the Air Force, 2000), 3. Also see Maj Gen John L. Barry and Col Darrell L. Herriges, "Aerospace Integration, Not Separation," *Aerospace Power Journal* 14 no. 2 (Summer 2000): 42.
[122] The Army, Navy and Marines in all theaters need space support at least as much as the Air Force.

in view to their theater.[123] Doing so would result in friction caused by warfighters in

adjacent theaters competing for the same space assets without a centralized mechanism to

eliminate duplication or prioritize limited resources. Also, allowing in-theater control of

satellites in low, medium, or highly elliptical orbits would likely degrade the readiness of

those satellites to perform their missions on the opposite side of the globe.[124] This would

be a repeat of North Africa.

Conflict occurs over the issue of who will exercise operational control of space

forces.[125] Current doctrine is a victim of editors who have simply replaced the word *air*

with *aerospace* or *air and space* in an honest attempt to be inclusive.[126] Unfortunately,

doing so "force fits" the attributes of airpower to spacepower and the theater mindset of

airmen to space professionals.[127] This is an endless source of confusion.

In its zest for aerospace integration, the Air Force now refers to its various theater air

operations centers (AOCs) as "aerospace" operations centers. This signals an increasing

effort by the Air Force to coordinate and control space support and counterspace

operations for the entire theater.[128] This is a job arguably best left to the combined or

joint staff because it involves coordinating space support for all theater components from

a CONUS-based unified command (USSPACECOM). The role of the handful of space

---

[123] "Penny packets," a term used to describe the almost worthless value of airpower when divided amongst ground units in Arthur Coningham's, "Development of Tactical Air Forces," *Journal of the Royal United Services Institution* (May 1946), Vol 91, 215.

[124] Apparently, the Joint Chiefs have complained because a new series of spy satellites that form part of the Future Imagery Architecture cannot take collection commands from commanders in the field. This would be a serious usurping of the centralized control principle. See Jeremy Singer, "Joint Chiefs Dissatisfied With New Series of Spy Satellites," *Space News* (5 February 2001): 1.

[125] AFDD 2-2 (Draft) *Space Operations*, (September 2000), 7-8.

[126] Lt Col Peter Hays and Dr Karl Mueller, "Boldly Going—Where? Aerospace Integration, The Space Commission, and the Air Force's Vision for Space," *Aerospace Power Journal* (Spring 2001): 36-37.

[127] Col Kenneth A. Myers and Lt Col John G. Tockston, "Real Tenets of Military Space Doctrine," *Airpower Journal* 2, no. 4 (Winter 1988): 54–68.

[128] AFDD 2, *Organization and Employment of Aerospace Power*, 17 February 2000, 34.

professionals inside each AOC is to advise the air component commander on space operations and spend the rest of their time coordinating with space centers in the CONUS. It is questionable whether an airman (or anyone else) can readily switch between a theater perspective when employing airpower and a global perspective when employing spacepower—the example of Army leaders failing to listen to their airmen advisors in North Africa looms.

When discussing where to centralize the control of space forces, the importance of the operator's perspective or mindset cannot be overemphasized. General Wilbur Creech, former commander of Tactical Air Command, credits the success of the Airland Battle Doctrine to General Donn Starry, Commander of the US Army's Training and Doctrine Center in the 1980s. Creech claims that Starry "saw combat in a much broader perspective than the traditional Army."[129] It was the first time that the Army recognized a theater perspective beyond the corps area, acknowledging the legitimacy of the airman's mindset.[130] Creech asserts that this mutual understanding laid the foundation for modern joint operations as later proven in Desert Storm.[131]

This lesson of mutual understanding and joint partnership appears lost on airmen who cannot appreciate the necessary differences between the theater mindset of airmen and the global mindset of space professionals. The roadblock is the insistence that air and space form a single operational medium called aerospace, and the denial that the effects that spacepower shares with landpower and seapower are just as great as those shared with airpower.

---

[129] Gen Wilber L. Creech, transcript of interview by Hugh N. Ahmann, June 1992, 221, US Air Force Historical Research Agency, Maxwell Air Force Base, Ala.
[130] Ibid., 222.
[131] Ibid., 223.

Just as the expanded mindset of airmen drove their need for centralized theater level control of airpower by an airman, so too, the further expanded mindset of space professionals drives their insistence that spacepower must be centrally controlled by a space professional.

#5 Spacepower is a Coercive Force

The photoreconnaissance satellite is one of the most important military technological developments of this century, along with radar and the atomic bomb. Without it, the history of this century would be very different. Indeed, without it history might well have ceased.

—Jeffrey Richelson, 1990.

We are entering an era—if we have not already entered it—when the use of space will exert such influence on human affairs that no nation can be regarded as a world power or remain a world power unless it possesses significant space capabilities.

—General Robert T. Herres

On 11 May 1998, India surprised the world by detonating nuclear test devices. The international diplomatic response was immediate and quite negative. The event prompted headlines that read "Pakistan Feels Let Down by US Spy Satellites," and "India Tricks US Satellites."[132] For Pakistan's part, they were counting on the US to provide advanced warning, presumably so their government could take whatever actions they might deem necessary for state security.[133] Indian officials later revealed that they managed to conceal the tests from US satellites by conducting nuclear tests "when sandstorms

---

[132] Tara Shankar Sahay, "Pakistan Feels Let Down by US Spy Satellites," *Rediff On the Net*, 13 May 1998, n.p.; on-line, Internet 17 April 2001, available from http://www.rediff.com/news/1998/may/13spy.htm, and Krishnan Gurswamy, "India Tricks US Satellites," *Associated Press*, 19 May 1998, n.p.; on-line, Internet 17 April 2001, available from http://abcnews.go.com/sections/world/DailyNews/india980519_nukes.html.
[133] Sahay, 1.

normally swept across the Thar Desert and intense heat could disrupt surveillance sensors. Activity was also timed around the flights of spy satellites."[134]

American spacepower clearly failed to deter India from conducting nuclear testing, but the presence of surveillance and reconnaissance satellites did coerce India into changing the pattern of its behavior. The threat of satellite detection compelled them to work around the gaps in satellite coverage. This begs the question, if the US had persistent coverage of India with spy satellites, would this have deterred India from taking actions leading to its joining the league of nuclear nations? Unfortunately, this is counterfactual. There is no way to answer this question. It remains that "coercive spacepower is not a foolproof means of bringing about a desired change in adversary behavior, no more so than coercive air, land, and seapower."[135]

In *Arms and Influence*, author Thomas C. Shelling describes coercion as "finding a bargain, arranging for [an adversary] to be better off doing what we want—worse off not doing what we want—when he takes the threatened penalty into account."[136] Robert Pape simplifies this idea somewhat when he describes coercion in as "efforts to change the behavior of a state by manipulating costs and benefits."[137] Coercion can take two basic forms. The first is deterrence, which seeks to preserve the status quo by discouraging an opponent from changing his behavior. The second is compellence,

[134] Gurswamy, 1.
[135] Maj Christopher J. Kinnan, "Coercive Spacepower: Enabling a "Virtual Boots on the Ground" National Security Strategy for the 21st Century." (end of course paper: School of Advanced Airpower Studies, 15 November 2000), 2.
[136] Thomas C. Schelling, *Arms and Influence* (New Haven: Yale University Press, 1966), 4.
[137] Robert A. Pape. *Bombing to Win: Air Power and Coercion in War*, (Ithica: Cornell University Press, 1996), 4.

which is a wide range of strategies that may include any combination of national instruments of power to force an opponent into accepting the terms of the coercer.[138]

Traditionally, spacepower's claim as a coercive force rests on the deterrent value of surveillance and reconnaissance satellites. Surveillance satellites that detect missile launch indications are an integrated part of America's nuclear deterrent architecture. The early detection capability they provide makes it possible for the US to formulate retaliation strategies that may include launching its ICBM force on warning, instead of absorbing an adversary's first strike. This may negate the value of a first strike designed to eliminate America's ICBM force in their silos. This form of coercion raises the cost and risk for any adversary who would attack the US.

Another deterrent value of spacepower is its asset's ability to serve as a national technical means of verification (NTMV) of arms control agreements. The Strategic Arms Reduction Talk (START) Treaties prescribe the method of demolishing bombers and missile silos specified in ways that facilitate satellite verification. The frequent inspection of the demolished equipment by satellites deters either state from attempting to break the treaty. This also raises the cost and risk for any adversary who would fail to comply with arms control agreements.

Some may argue that spacepower itself is not a deterrent force, that it is the reaction of politicians and diplomats armed with the information gleaned from satellites that is the real deterrent force at work. True, deterrence is based on the presumption that undesirable actions will meet with undesirable consequences, but it remains that the presence of the surveillance and reconnaissance satellites themselves shape a potential adversary's behavior. India's effort to work around spacepower's operating limitations is

---

[138] Schelling, 69-78.

71

evidence. India had no way of knowing whether or not the reconnaissance satellites passing overhead were operational. Still, they recognized the presence of the satellites themselves as a threat and reacted accordingly.

In the past spacepower's coercive force rested upon the deterrent value of its assets, but increasingly it is contributing to compellence efforts as well. The decade of the 1990's ushered with it an evolution of the spacepower mission. Global Positioning System (GPS), only a budding technology during Operation Desert Storm, enabled the US Army to maneuver flawlessly through the desert of the Middle East. However, with the advent of technologies such as Joint Direct Attack Munitions (JDAM), spacepower began to play a more direct role in the force application (compellence) mission.[139] In this sense, spacepower is moving from an information role to being an integral part of the "sensor-to-shooter kill chains."[140] This reality is beginning to gray the lines between force enhancement and force application, and deterrence and compellence, though the tide is still probably closer to enhancement and deterrence.[141]

Looking ahead, it appears spacepower will continue to grow as a coercive force. Improved intelligence, surveillance, and reconnaissance satellite networks will improve spacepower's deterrence potential. This will ultimately close the gaps in our coverage that India found and exploited. Eventually space will be weaponized, and doing so will bolster both deterrence and compellence capabilities. Obviously, both of these

---

[139] Maj John McMullen, "Spacepower as a Coercive Force," School of Advanced Airpower Studies, Maxwell AFB, Ala., 16 April 2001.
[140] Barry D. Watts, *The Military Use of Space: A Diagnostic Assessment,"* Center for Strategic and Budgetary Assessments Report (Washington D.C.: Center for Strategic and Budgetary Assessments, February 2001), 13.
[141] Ibid., 43.

developments will have far reaching geopolitical and strategic impacts, but the US is carefully assessing these potentialities.

Major General Lance Smith, commander of the Air Force Doctrine Center, was the senior leader in the recent "Schriever 2001" space wargame held at Colorado's Schriever Air Force Base in January 2001.[142] In the scenario, he played the commander of a bolstered US Space Force that had at its disposal a wide variety of potential space-based improvements that may become reality by the 2017 timeframe in which the wargame was set. In an interview to the press during the wargame, he explained that spacepower is a coercive force that may offer the ability to suppress a potential adversary without resorting to actual force. "Today, we use lethal options when we want to demonstrate resolve, such as bombing nonessential targets," Smith said. "Space could allow us to demonstrate to an adversary how we can hurt him militarily and economically without actually carrying it out and force him to negotiate. It may allow a face-saving dimension for the enemy which we hadn't thought about before."[143]

William Scott, a writer for *Aviation Week and Space Technology*, commented on these same wargames, "Having a robust space force could actually promote global stability, effectively deter a potential aggressor, and avoid armed conflict. Simply being able to constantly monitor the buildup of an adversaries forces, then publicly display imagery of them, can be a major deterrent."[144] The US demonstrated this at the United

---

[142] Maj Gen Smith is a career fighter pilot and an avid supporter of the aerospace integration concept. Maj Gen Lance Smith, Maxwell AFB, Ala., interview with author, 9 March 2001.

[143] Mike Patty, "Air Force Conducts Virtual Space War," Denver Rocky Mountain News, 26 January 2001, on-line, Internet 29 January 2001, available from https://ca.dtic.mil/cgibin/ebird?doc_url=/Jan2001/s20010129conducts.htm.

[144] William B. Scott, "Wargame: 'Space' Can Deter, Defuse Crises," *Aviation Week and Space Technology*, 5 February 2001, 40.

Nations when it revealed imagery of Soviet missiles in Cuba during the Cuban missile crisis.

The purpose of spacepower is to achieve goals and objectives though the control and exploitation of the space environment. This is possible simply because space assets collect and disseminate information that decision makers can exploit. On another level, the presence of space sensors deters some actors from certain forms of behavior. Increasingly, spacepower is inside the closed sensor-to-shooter loop and eventually weapons will migrate to space. All of these factors indicate that spacepower is a coercive force.

#6 Commercial Space Assets Make All Actors Space Powers

Mercenaries are worth three men; one in our army, one who is not in our enemy's army, and one of our citizens that can remain at work and pay taxes.

—Frederick the Great

Commercial satellites are worth three military satellites; one in our service, one that is not in our enemy's service, and one less satellite program to pay for.

—Coyote's Space Corollary to Frederick's Mercenaries

If you have a credit card and access to a telephone or the Internet, you can become your own space power.[145] Like the condottieri of the Italian renaissance, several companies are ready to sell commercial space products to anyone who can meet their

---

[145] Gen Richard B. Myers, "Space Superiority is Fleeting," *Aviations Week and Space Technology*, 1 January 2000, on-line, Internet 15 July 2000, available from http://www.peterson.af.mil/usspace/avweek-gen%20myers.htm.

price.[146] The types of services for sale include photographic imagery down to 1.0-meter resolution (soon to be 0.5-meter resolution), infrared detection, radar scanning, communications, Global Positioning System (GPS) receivers, and access to refined weather data.[147] Just a few years ago, these capabilities were the exclusive privilege of the superpowers, but today they can be yours if the price is right.

This is becoming a major issue in modern competition between actors. Among the findings of the recent Space Commission was that "small nations, groups or even individuals can acquire from commercial sources imagery of targets on Earth and in space. They can acquire accurate timing and navigation date and critical weather information generated by government-owned satellites. Improved command and control capabilities are available using commercial communications capabilities. Even launch capabilities can be contracted for with legitimate companies."[148] The report goes on to note, "Commercial satellite ground communications equipment has electronic jamming capabilities that can easily be used to disrupt the functions of space satellites."[149] Perhaps an even greater concern is the fact that the Aviaconversia Company in "Russia is [openly] marketing a handheld GPS jamming system (figure 3)."[150] A 4-Watt version of the device was displayed at the Moscow Airshow in September 1997, with a price tag of less than $4,000.[151] Such a device threatens to jam GPS signals used to aid the navigation of aircraft and several other weapons systems at ranges over 100 miles. It may also deny

---

[146] The condottieri were leaders of mercenary bands or companies that hired out their services as soldiers of fortune between the 14th and 16th centuries in the Northern Italian provinces.
[147] Frank G. Klotz, *Space, Commerce, and National Security* (New York: Council on Foreign Relations Press, 1998), 8.
[148] Space Commission Report, 19.
[149] Ibid.
[150] Ibid., 20.
[151] Charles Seife, "Where Am I," Info-Sec.Com, on-line, Internet, 10 April 2001, available from http://www.info-sec.com/denial/denial_012298a.html-ssi.

access to the GPS timing signal that modern military and commercial digital networks increasingly rely upon to integrate and synchronize communications and information.

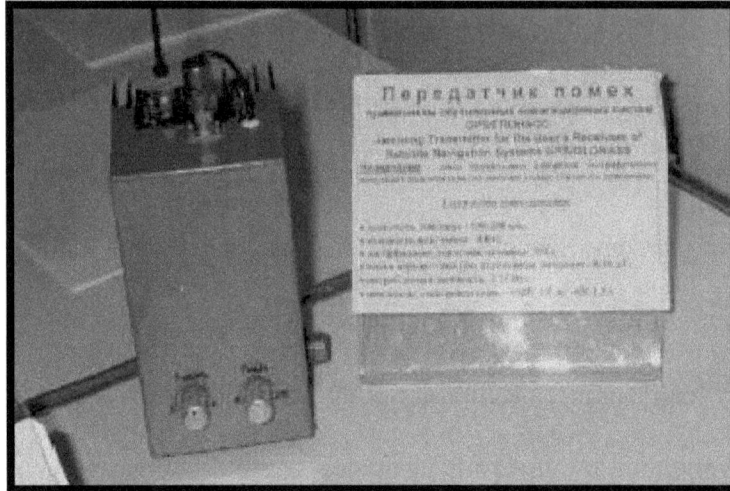

Figure 3. Aviaconversia Commercial GPS Jammer

Source: Attributed to the National Air Intelligence Center by Charles Seife, "Where Am I," Info-Sec.Com, on-line, Internet, 10 April 2001, available from http://www.info-sec.com/denial/denial_012298a.html-ssi.

Commercial satellite service vendors are becoming modern-day mercenaries and the United States is helping to make them so. During Desert Storm, more than 90 percent of all the long distance communications used by American forces went through space.[152] Commercial satellites provided nearly 25 percent of this service.[153] At the same time, the US Air Force purchased more than 100 SPOT images of downtown Baghdad.[154] The American trend of using commercially derived space products to augment its own space capabilities continues to grow as the nation seeks greater cost savings by leveraging against capabilities available in the market place. For example, following Allied Force,

---

[152] Klotz, 7.

[153] Bob Preston, *Plowshares and Power: The Military Use of Civil Space* (Washington, D.C.: National University Press, 1994), 132. Cited by Klotz, 9.

[154] SPOT is an acronym for *Satellite Pour l'Observation de la Terre*. It is a French commercial venture. From Klotz, 9.

USSPACECOM estimated that 80 percent of the spaceborne communications used in the operation traveled on commercial satellites.[155] As Lieutenant Commander J. Todd Black stated in a *Naval War College Review* article, "Commercial satellite systems are quickly becoming indispensable to the US military, and they are almost certainly growing more useful to potential enemy military, paramilitary, terrorist, and other unconventional forces."[156]

The consequences of facing an adversary armed with the types of information derived from commercial space systems would be profound. If, for example, Iraq had had access to imagery and navigational data, the coalition's victory would not have been as easy. They might have unmasked preparations for the ground force's "left hook" into Iraq.[157] They certainly weren't expecting an attack out of their uncharted desert because the enemy knew, as General Norman Schwartkopf points out, "whenever his own forces went out there, they always got lost."[158] GPS changes all of that. Today's adversary will use GPS to his own advantage.

There is a unique dilemma brought about by the sale of commercial space products. It is likely that *both* parties in a dispute will purchase satellite services not only from the same company, but also derived from the same satellite. This actually happened during Operation Allied Force, wherein the Serbians leased communications bandwidth from the EUTELSAT Corporation, a corporation that operates a fleet of communications satellites. Ironically, NATO nations were sharing a particular satellite with the Serbs. Eventually,

---

[155] Watts, 41.

[156] LTC J. Todd Black, "Commercial Satellites: Future Threats or Allies?" *Naval War College Review* 52, no. 1 (Winter 1999), on-line, Internet 10 April 2001, available from http://www.nwc.navy.mil/press/Review/1999/winter/art5-w99.htm.

[157] Klotz, 25.

[158] James Oberg, "Spying for Dummies," Spectrum Magazine (November 1999) on-line, Internet 10 April 2001, available from http://jamesoberg.com/articles/spy/.

diplomats raised this issue with the corporation's Assembly of Parties, who voted to suspend service to the Serbians until a later date.[159] This raises serious legal and military issues. How should the parties in a conflict treat a common service provider? What legal obligations does the service provider have to the warring parties? There is nothing in space law to establish a suitable precedent.

Earlier this year, a military exercise at Schriever Air Force Base, Colorado, examined the problems of using commercial space products in warfare. The war game was hypothetically set in the year 2017 with country "Red" massing forces on its boarder for possible attack against its smaller neighbor, "Brown." Brown asked "Blue" (presumably the United States) for help.

> On Day 3 of the game, privately owned foreign satellites became a key issue. The Blue side asked the foreign firms not to provide services to Red. In response, Red tried to buy up all available services to constrain the US military, which relies heavily on commercial space satellites for many of its communications [and other capabilities]. Red offered to pay far more that is customary. Blue then said it would top Red's offer. The eight people playing the foreign firms responded that they would honor their contracts, which left Blue worried and unhappy. Robert Hegstrom, the games director, concluded that "dealing with third party commercial providers is going to be a priority for USCINCSPACE."[160]

There are still no good solutions to the military problems presented by commercial satellites. In "The Opening Skies: Third-Party Imaging Satellites and US Security," author Ann M. Florini states that militaries have three basic options in dealing with commercial imaging satellites, although Black claims these options also apply to other satellite types as well. The first option is to adopt a *free market approach*, simply

---

[159] "Transcript: NATO Briefing On Operation Allied Force, 28 May 1999," on-line, Internet 10 April 2001, available from http://www.usembassy.it/file9905/alia/99052814.htm.

[160] Thomas E. Ricks, "US Air Force Prepares Itself to Do Battle in Outer Space," *Washington Post Service*, 30 January 2001, on-line, Internet 31 January 2001, available from http://www.ith.com/cgi-bin/generic.cgi?template=articleprint.tmplh&ArticleId=9116.

attempting to outbid one's opponent or beat him to the punch by obtaining contracts early. The second option is to attempt to *negotiate agreed restraints* in the marketplace and international assemblies. The final approach is taking *direct countermeasures* against satellites, their data-gathering, or their ground systems.[161] It is likely that some states will attempt a combination of all three approaches, but there is no telling what courses of action the various actors around the globe will ultimately pursue.

The problems presented by commercial space assets are still so new that no one has yet decided how best to handle them. While advanced nations use commercial satellites to augment their own intrinsic space assets, new actors now have access to information that levels the informational playing field considerably. There is no doubt that commercial satellites are a new form of mercenary that the various actors will covet. The bottom line is that commercial space assets make all actors space powers.

#7 Spacepower Assets Form a National Center of Gravity

Space is becoming an economic center of gravity for the United States and may well become such for other nations.

—Gen Howell M. Estes, III, USCINCSPACE, 1997

Space...is increasingly at the center of our national and economic security...[S]pace is not just a military, but also an economic center of gravity, and unarguably, a vital national interest.

—Gen Richard Myers, USCINCSPACE, 1999

---

[161] Ann M. Forini, "The Opening Skies: Third-Party Imaging Satellites and US Security," *International Security*, Fall 1998, 103-123, and Black, 5.

In *On War*, Clausewitz describes a center of gravity as "the hub of all power and movement, on which everything depends."[162] It is therefore a source of strength and at the same time a vulnerability requiring protection. He claims it is "always found where the mass is concentrated most densely."[163] However, Clausewitz concedes that an enemy may possess several centers of gravity. "It is therefore a major act of strategic judgment to distinguish these centers of gravity in the enemy's forces and to identify their spheres of effectiveness."[164] By identifying where their spheres of effectiveness overlap, it becomes possible to trace multiple centers of gravity back to a single one.[165] If it is possible to reduce the enemy's centers of gravity to one, "it represents the most effective target for a blow."[166] In essence, Clausewitz is describing the logic behind nodal analysis and effects based targeting.

Spacepower assets (satellites, ground stations, and data links) are not *the* national center of gravity, but are *a* center of gravity, which have spheres of effectiveness that overlap in sectors of civil, commercial, military, and intelligence activity. The relative value of spacepower assets depends on how much an actor uses them and for what. For a small state such as Burundi, spacepower assets do not form a significant center of gravity, although they are a potential space power by virtue of access to commercial space products. However, as Barry Watts points out, "American requirements for global power projection suggest that the Unites States is more dependent on space systems than other

---

[162] Carl von Clausewitz, On War, ed. and trans. Michael Howard and Peter Paret (Princeton, N.J.:Princeton University Press, 1984), 595.
[163] Ibid., 485.
[164] Ibid., 486.
[165] Ibid., 619.
[166] Ibid., 485.

countries."[167]   In effect, spacepower assets are a larger center of gravity for the United States than other countries—at the turn of the 21st century.   Each sector deserves separate consideration as a center of gravity, but also with regard to the overlap of its sphere of effectiveness with other space and terrestrial interests.

Civil Space Sector.   The civil space sector does much more than manned space flight and space exploration for scientific curiosity.   It is a driving engine of scientific research and discovery.   Many of the improvements in telecommunications, microelectronics, computing, and machine engineering that we enjoy today have roots in space research sponsored by programs such as Mercury, Gemini, Apollo, and others. Scientific and technological spin-offs from the civil space sector not only fuel high-tech commercial industries, but it also generates popular enthusiasm and promotes the types of math and science educational programs that fuel a high-tech society.   Centers of research for the civil space sector represent highly lucrative espionage targets, and as the Challenger disaster in 1986 demonstrated, fatal glitches with the manned space program can significantly stymie progress in this sector.

Commercial Space Sector.   The biggest change in spacefaring activities is the recent emergence and now dominance of the commercial space sector.   Whereas governments drove space activity in early days of space venturing, in the late 1990s commercial businesses launched more payloads into space than governments.   This trend will continue, with revenues exceeding $80 billion in 2000 and projected to more than triple in the next decade.[168]   While this is a drop in the bucket of the overall global

---

[167] Barry D. Watts, *The Military Use of Space: A diagnostic Assessment,"* Center for Strategic and Budgetary Assessments Report (Washington D.C.: Center for Strategic and Budgetary Assessments, February 2001), 1.
[168] Space Commission Report, 12-13.

economy in terms of dollars, it is far more important to consider what type of information the commercial space sector is collecting or moving and what capabilities are lost if these satellites are negated. Information carried on commercial satellites includes banking information, credit card authorization networks, video feeds for cable and broadcast feeds, cellular telephone networks, pager networks, communications networks, and corporate communications systems rely heavily on the commercial space sector.[169] In addition, the armed forces increasingly rely on commercial satellites for services as gap fillers for their own space systems, or as backups in case a military satellite fails. In sum, the commercial space sector is not only a profit maker, but more importantly, it provides an infrastructure for key pieces of governmental, military, and economic information. Of greatest concern is the relative importance of commercial satellites to society as a whole, which are relatively few in number, and poorly protected. This poses a highly lucrative target set to adversaries.

Military Space Sector. Political leaders and terrestrial forces rely increasingly on informational support from space in order to find, fix, track, target, and guide munitions against enemy targets. It is an overstatement to claim that terrestrial forces are dependent on space support, but it is fair to say that space support facilitates greater situational awareness in the battlespace and thereby increases the timing, tempo, and precision of friendly forces. Without space support, political leaders and terrestrial forces would have to operate more slowly using previous generations of technology. There is a growing union between the military and commercial space sectors. The armed forces increasing rely on commercial services in lieu of fielding dedicated military systems. In part, this propels the commercial sector forward by providing capital, but it limits the military to

---

[169] Ibid., 22.

satisfying its needs with generic services instead of highly specialized capabilities. Military systems also assist the commercial and civil sectors as well. The US Air Force, for example, provides global access to the Global Positioning System, allowing the commercial market to profit by selling receivers to the civilian market. US Space command also provides space object tracking to all of the space sectors. Still, the satellites operated by the military represent an asymmetric advantage over adversaries who lack similar capabilities. Any adversary who bases their strategy on leveling the technological playing field with the must consider negating military satellites.

Intelligence Space Sector. It is sometimes difficult to separate the military and intelligence space sectors from each other in terms of operations since the two work hand-in-hand to provide capabilities to political leaders to assist in achieving the goals of policy. The separation came about during the Cold War wherein the Eisenhower Administration established the National Reconnaissance Office (NRO), a "black" organization shrouded in secrecy until the 1990s, to operate spy satellites to peer behind the Iron Curtain. Conversely, the military became the repository for openly acknowledged "white" space systems, although the actual missions and capabilities of many military systems remained classified, they did not intrude on the NRO's intelligence gathering mission. Today, the NRO employs its satellites as a national technical means of verification (NTMV) for treaty compliance, as well as providing critical intelligence, surveillance, and reconnaissance (ISR) support to political leaders and military forces alike. Like the military, the intelligence sector increasingly buys services from the commercial sector to fill its imaging and other sensing and communications requirements. Satellites of the intelligence space sector, in concert with surveillance platforms in the military space sector, act as global sentries. Together, they keep a watch on potential trouble areas and signal political and military leaders whenever trouble is afoot. These satellites are likely targets early in a conflict for an adversary who seeks to preserve secrecy or gain the element of surprise.

There is tremendous overlap between the sectors. Satellite systems often share the same intellectual and industrial base, the same launch facilities, the same control network, and sometimes the same portions of the electromagnetic spectrum. These present highly lucrative targets to an adversary. When considering a single asset, such as a communications satellite, it may carry signals of hundreds or thousands of users all at once. Any degradation to a satellite supporting multiple users has an immediate global effect that reduces a state's ability to operate abroad. However, access to similar commercial space assets adds capacity to a state's spacepower architecture from a third party, which suggests there may always be some access to space services even in a war that exacts high attrition on a state's proprietary spacepower assets.

The importance of spacepower to the US is widely recognized by political and military leaders alike. Increasingly, the use of space is viewed not only as an inherent force multiplier, but also as a vulnerability that must be addressed.

> Soon after assuming command of the US Space Command, General Estes noted that "we are the world's most successful space-fairing nation…One of the major reasons the United States holds its current position in today's league of nations. But, we are also the world's most space-dependent nation thereby making us vulnerable to hostile groups or powers seeking to disrupt our access to, and use of, space. For this reason, it is vital to our national security that we protect and safeguard our interests in space. The ability of our potential adversaries to affect our advantage in space is

growing. We, in military space, are just now beginning to consider and deal with these threats."[170]

Considering only military spacepower, the US armed forces currently use spacepower assets for two primary purposes, first, to improve the situational awareness of its forces, and second, as a means of command, control, and communications with its forces. The US essentially exploits spacepower assets as a permanent informational infrastructure that is globally available to friendly forces. This allows friendly forces to operate on interior lines of information around the globe.

No claim is made that US military forces are otherwise neutered without space support. Terrestrial forces can still fight without space support. However, lacking space support will inarguably increase the fog, friction, and overall costs of military operations. Terminating friendly force's access to space will force them to move information on exterior lines.

Likewise, no claim is made that an attack that totally neutralizes America's spacepower assets will, of itself, be decisive. However, if an adversary can increase the fog, friction, and costs of American operations by striking some or all of its spacepower assets, it is possible that this may turn the tide of battle against US forces. If the enemy is powerful enough, or smart enough, this may set up the conditions to defeat the US.

As space products and services become ever more interwoven with American politics, economics, culture, and security, they become increasingly lucrative targets for potential adversaries.[171] This is rooted in the basic logic spelled out by Clausewitz.

---

[170] As quoted in Maj David W. Ziegler, "Safe Havens, Military Strategy and Space Sanctuary Thought," (Maxwell AFB, Ala., June 1998), 22.
[171] *Long Range Plan: Implementing USSPACECOM Vision for 2020* (Peterson AFB, Col.: U.S. Space Command, Director of Plans, April 1998) n.p.; on-line, Internet 11 April 2001, available from http://www.peterson.af.mil/usspace/LRP/cover.htm.

When extrapolated to the 21st century, we realize that spacepower assets are a national center of gravity.

#8 Space Control is Not Optional

The United States must win and maintain the capability to control space in order to assure the progress and pre-eminence of the free nations. If liberty and freedom are to remain in the world, the United States and its allies must be in position to control space.

—Gen Thomas D. White
Air Force Chief of Staff

Control of space means control of the world, far more certainly, far more totally than any control that has been achieved by weapons or troops of occupation. Space is the ultimate position, the position of total control over Earth.

—Lyndon Baines
Johnson

Unimpeded access to and use of space is and will remain a vital national interest.

—Annual Defense Review, 2001

In the Second World War, seapower held vastly different strategic significance for the Allies than it did for Germany. The Allies depended upon freedom of the sea to logistically connect the alliance and to take the war to their continental foe. Germany, on the other hand, had little need for the sea, other than to deny its use to its enemies.[172] Accordingly, Germany dedicated substantial resources to build submarine wolf packs to interdict allied shipping in the Battle of the Atlantic. Germany's early successes cost the Allies precious time and resources. Prime Minister Winston Churchill later commented "The only thing that truly scared me during the Second World War was the Battle of the

Atlantic."[173] Fortunately, Germany was unable to capitalize on the increased fog and friction their U-boats caused among the Allies. In the end, the Allies gained control of the sea and ultimately defeated the Axis powers.

Today, spacepower holds vastly different strategic significance for the United States than it does for other countries. "At the dawn of the 21st century, the preeminent user of near-Earth space for military purposes is the United States...[which] is currently far ahead of any other nation in the capability to exploit orbital systems for the enhancement of terrestrial military operations."[174] American forces use space systems as a global informational infrastructure that friendly forces stationed or deployed worldwide can plug into in order to receive services that increase situational awareness, improve precision engagement, and expedite command and control. Spacepower essentially increases the timing, tempo, and precision of American operations beyond the ability of adversaries to respond in kind. This translates into a very compelling incentive for adversaries to counter American space lines of communications, whether or not they themselves are spacefaring nations.[175]

To date, the U.S. has enjoyed complete freedom to capitalize on space systems to bolster its national power, while become increasingly reliant on space systems for its economic and military success. In fact, the Space Commission Report states, "The US is more dependent on space than any other nation," but the ability for political and military

---

[172] Colin S. Gray and John B. Sheldon, "Spacepower and the Revolution in Military Affairs: A Glass Half Full?" *Airpower Journal* 8, no. 3, (Fall 1999): 31.

[173] As cited by "Harry Tate's Navy: The Royal Naval Patrol Service," British-Forces.com, on-line, Internet 12 May 2001, available from http://british-forces.com/links_recom/navy.html.

[174] Watts, 1.

[175] Watts, 12.

leaders to take unimpeded access to space-based capabilites may rapidly be coming to an

end.[176] As General Richard B. Myers, USCINCSPACE, noted in January 2001:

> We can't be deceived by the fact that we enjoyed space dominance in
> Kosovo and in the Gulf War. We controlled the high ground, not because
> of superior technologies or strategies, but because our adversaries didn't
> use space. We gained space superiority by default; this was our bye
> round, and a key take-away is that the whole world took notice. Just as
> Milosevic modified his air defenses to try and deny our air superiority,
> others will modify their forces to try and deny our space superiority."[177]

Other nations did take notice. In July 2000, a Chinese news agency reported that its

military is developing the means to defeat American space-based systems, stating that

"for countries that could never win a war by using the method of tanks and planes,

attacking the U.S. space systems may be an irresistible and most tempting choice..."[178]

Subsequently, China revealed that it is developing tiny "parasitic satellites" to be used as

anti-satellite weapons, claiming that this development will "drastically change the

Chinese-American military balance so that the U.S. would not intervene easily in the

event of a conflict in the Taiwan Strait..."[179] Despite mounting evidence that more

nations intend to challenge America's space dominance, American space control efforts

have focused heavily on tracking objects on orbit (space surveillance) and not on

defending friendly space assets or negating adversary space threats. There is a growing

concern that without a concerted and balanced space control effort that America might be

setting itself up for a "Space Pearl Harbor."[180]

---

[176] Space Commission Report, 18.
[177] Gen Richard B. Myers, "Space Superiority is Fleeting," *Aviation Week and Space Technology*, 1 January 2001, n.p.; on-line, Internet 15 July 2000, available from http://www.peterson.af.mil/usspace/avweek-gen%20myers.htm.
[178] Space Commission Report, 22-23.
[179] Cheng Ho, "China Eyes Anti-Satellite System," *Space Daily*, 8 January 2000, n.p.; on-line, Internet 8 January 2001, available from http://www.spacer.com/news/china-01c.html.
[180] This is actually asserted seven separate times in the report's executive summary and main body. Space Commission Report, viii, xiii, xiv, xv, 22 (twice), 25.

There is a new awakening to the space threat among the Washington insiders. In the 2000 *Annual Defense Review* (ADR), the Secretary of Defense merely characterized space control as posing a "significant challenge to U.S. defense strategy.[181] However, in the 2001 *ADR,* the Secretary of Defense raised the relative importance of space control to a "vital national security interest" and succinctly described space control this way:

> The ability of the United States to access and utilize space is a vital national security interest because many of the activities conducted in space are critical to its national security and economic well-being. Potential adversaries may target and attack U.S., allied, and commercial space assets during crisis or conflict as an asymmetric means to counter or reduce U.S. military operational effectiveness, intelligence capabilities, economic and societal posture, and national will. Therefore, ensuring the freedom of space and protecting U.S. national security interests in space are priorities for the Department.
>
> The mission of space control is to ensure the freedom of action in space for the United States and its allies and, when directed, deny an adversary freedom of action in space. The space control mission area includes: the surveillance of space; the protection of U.S. and friendly space systems; the prevention of an adversary's ability to use space systems and services, the negation of adversary space systems and services; and supporting battle management, command, control, communications, and intelligence.[182]

What has changed to stimulate this new awareness, besides a change of presidents? The simple answer is congressional interest and new intelligence. In 1999 several congressional commissions were established to assess America's progress in space, all reporting (among their many findings) that the nation is increasingly dependent on spacepower and proportionately vulnerable to space attacks. One of these reports, The Space Commission Report, cited the following indications of the scope of America's potential vulnerabilities:

---

[181] Secretary of Defense, *Annual Defense Review*, 2000 (Washington, D.C.: Government Printing Office), 97 (henceforth ADR 2000).

- In 1998, the Galaxy IV satellite malfunctioned, shutting down 80 percent of U.S. pagers, as well as video feeds for cable and broadcast transmission, credit card authorization networks and corporate communications systems. To restore satellite service, satellites had to be moved and thousands of ground antennas had to be manually repositioned, which took weeks in some cases.

- In early 2000, the U.S. lost all information from a number of its satellites for three hours when computers in ground stations malfunctioned.

- Hackers are routinely probing DoD networks and computers. The U.S. Space Command's Joint Task Force for Computer Network Defense reported that detected probes and scans are increasing, access to hacking tools is becoming easier, and hacking techniques are becoming more sophisticated. In 1999, the number of detected probes and scans against DoD systems was just over 22,000; in the first eleven months of 2000, the number had grown to 26,500.

- If the GPS system were to experience widespread failure or disruption, the impact could be serious. Loss of GPS timing could disable police, fire and ambulance communications around the world; disrupt the global banking and financial system, which depends on GPS timing to keep worldwide financial centers connected; and interrupt the operation of electric power distribution systems.[183]

The specter was raised on 7 February 2001, when Vice Admiral Thomas R. Wilson, Director of the Defense Intelligence Agency, testified openly before the Senate Intelligence Committee, "A number of countries are interested in or experimenting with a variety of technologies that could be used to develop counterspace capabilities."[184] He added, "China and Russia have across-the-board programs underway, and other smaller states and nonstate entities are pursuing more limited—though potentially effective—

---

[182] Secretary of Defense, *Annual Defense Review*, 2001 (Washington, D.C.: U.S. Government Printing Office), 128 (henceforth ADR 2001).

[183] Space Commission Report, 22-23.

[184] Vice Adm. Thomas R. Wilson, "Global Threats and Challenges Through 2015," (Statement for the Record Senate Select Committee on Intelligence, 7 February 2001), 14, on-line, Internet 12 April 2001, available from http://www.dialumni.org/images/dr_testimony.pdf

approaches.[185] It is often said that "freedom isn't free." Now it appears that freedom in space is not going to be free much longer either.

Faced with growing economic and military dependency in space, the U.S. cannot afford to leave its space assets undefended, nor leave itself open to attacks from space. Just as mankind developed navies to protect its investments transiting the sea, so too the U.S. is compelled to develop more robust space forces to protect its interests in space. This challenges the notion of "freedom in space," and "space for peaceful purposes." Given the long-standing international body of laws governing "freedom of the seas," and mankind's willing abandonment of such restraints in wartime, the future of space as a permanent peaceful sanctuary appears bleak.

Gray and Sheldon summarized the situation quite nicely in their article "Space Power and the Revolution in Military Affairs: A Glass Half Full," when they wrote:

> Space control is not an avoidable issue. It is not an optional extra. If the U.S. armed forces cannot secure and maintain space control then they will be unable to exploit space reliably, or reliably deny such exploitation to others. The U.S. ability to prevail in conflict would be severely harmed as a consequence. If you fail to achieve a healthy measure of space control in the larger of the possible wars of the [21st] century, you will lose.[186]

In future wars, the "Battle for Space" may be analogous to the Battle of the Atlantic. Preserving its space lines of communications is a vital national interest of the US during war and peace. The bottom line is that space control is not optional.

#9 Space Professionals Require Career-Long Specialization

We're going to solve our reconnaissance problem once and for all. Get on the horn and tell them to park a [reconnaissance satellite] directly above Baghdad.

---

[185] Ibid.
[186] Colin S. Gray and John B. Sheldon, "Spacepower and the Revolution in Military Affairs: A Glass Half Full?" *Airpower Journal* 8, no. 3, (Fall 1999): 36.

—Unnamed Commander of Joint Task Force-Southwest Asia

How many 'Gs' is that satellite pulling (pointing to sinusoidal peaks made by the ground track of a LEO satellite on a flat map projection)?

—Unnamed USCINCSPACE (a career fighter pilot) in his first week on the job

The Department of Defense must create a stronger military space culture, through focused career development, education and training, within which the space leaders of the future can be developed.

—Space Commission Report

Even though eighty-five percent of the space-related budget and personnel activity inside the DoD resides in the Air Force, flying aircraft and operating satellites are as different as night and day.[187] There is no common core competency between aircraft operators and spacecraft operators. One of the most important findings of the Space Commission is that "The Department of Defense [read as Air Force] is not yet on course to develop the space cadre the nation needs."[188] The Commission noted that few space professionals are in leadership positions in either Air Force or US Space Commands. Instead, the Air Force and DoD have installed senior officers who typically lack any space expertise whatsoever into space leadership roles.

"General Carl Spaatz once commented in exasperation that soldiers and sailors spoke solemnly about the years of experience that went into training a surface commander, thus making it impossible for outsiders to understand their arcane calling. Yet, they all felt perfectly capable of running an air force. That comment, echoed by American airmen for

---

[187] Space Commission Report, 55.
[188] Ibid., 42.

92

decades, was at the root of their calls for a separate air force."[189]   Many space

professionals believe that airmen place them in a similar predicament:

> In the Air Force pilot and Navy nuclear submarine career fields, military leaders have spent about 90 percent of their careers within their respective fields. In contrast, military leaders with little or no previous experience in space technology or operations often lead space organizations. A review by the [Space] Commission of over 150 personnel currently serving in key operational leadership positions showed that fewer than 20 percent of the flag officers in key space jobs come from space career backgrounds (see Figure 4). The remaining officers, drawn from pilot, air defense artillery, and Intercontinental Ballistic Missile (ICBM) career fields, on average spend 8 percent, or 2.5 years, of their careers in space or space related positions. Officers commanding space wings, groups, and squadrons fare only slightly better; about one-third of the officers have extensive space experience, while the remaining two-thirds averaged less than 4.5 years in space related positions (see Figure 5).[190]

---

[189] Col Phillip S. Meilinger, *Ten Propositions Regarding Air Power* (Washington, D.C.: Air Force History and Museums Program, 1995), 49.

[190] Space Commission Report, 43.

Figure 4. Flag Officers in Space Operations Positions

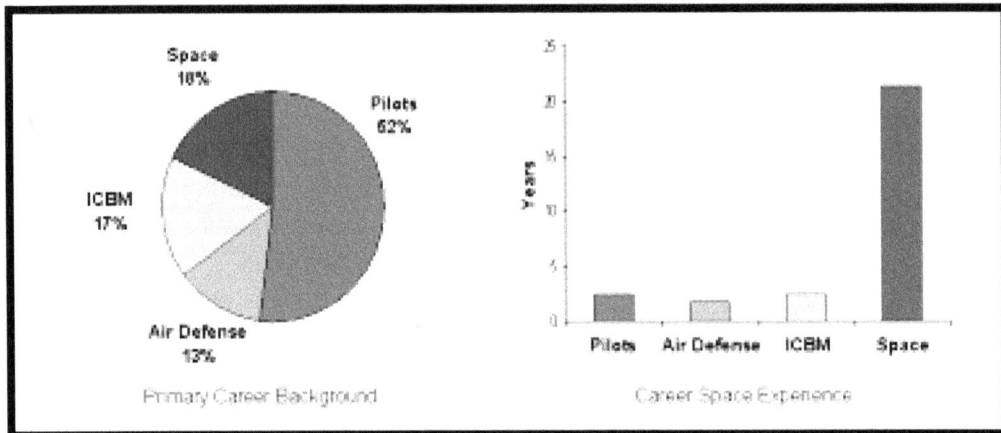

Source: Space Commission Report, 43.

Figure 5. Field Grade Officers in Space Operations Positions

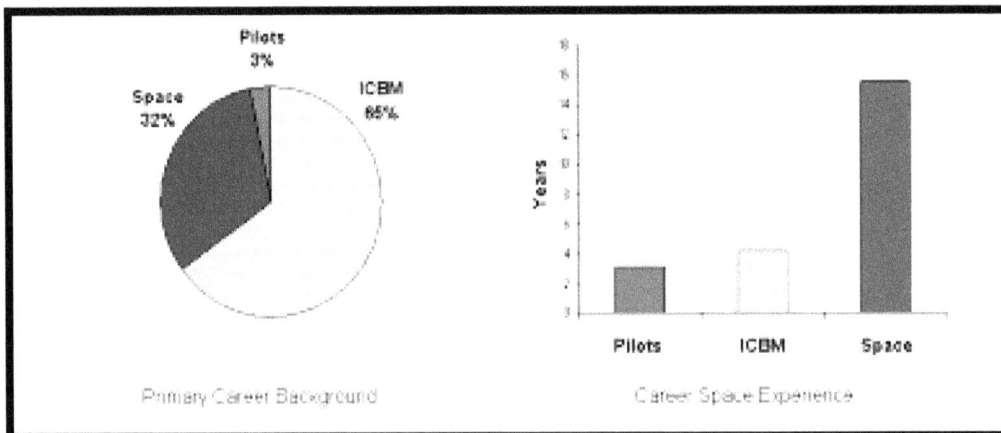

Source: Space Commission Report, 43.

In a paper prepared for the Space Commission, titled "Military Space Culture," Lt Col J. Kevin McLaughlin asserts "This keeps space organizations from reaching their potential. Today, many leaders of space organizations spend most of their assignments learning about space rather than leading."[191] The short tour length of senior officers,

---

[191] Lt Col J. Kevin McLaughlin, "Military Space Culture," Space Commission Background Papers, n.p.; on-line, Internet 15 April 2001, available from http://www.space.gov/commission/support-docs/article02/article02.html.

averaging less than eighteen months, exacerbates this problem.[192] As a result, the lexicon of space discussions seldom rises above the elementary level because the boss will not understand. Furthermore, senior leaders who lack space expertise cannot immediately grasp the contextual elements of problems or issues that may arise. Finally, keeping the boss straight places a heavy burden on the staff.[193]

Part of the problem may be the tendency inside the DoD to think of space as a mission, instead of a place where various mission are performed. Evidence of this tendency is found in the way the US Air Force Weapons School is organized. The school is comprised of eleven divisions. Each combat aircraft has its own division allowing its operators to develop highly specialized skills, but all things space related are combined into a single space division, forcing space operators to develop more generalized skills. Another example is Air Force Tactics, Techniques, and Procedures (AFTTP) Manual 3-1. Each aircraft has its own volume, but all of the various satellites and space systems are combined into a single volume.[194] In reality, each of the various missions performed in space requires a cadre of highly qualified and relatively specialized personnel to do the job right. This is significant because the disparity between space systems is as least as large as the disparity between aircraft types.

Currently, the space career field is currently divided into five major mission areas: (1) Satellite Command and Control, (2) Spacelift Operations, (3) Missile Operations, (4) Space Surveillance, and (5) Space Warning. McLaughlin claims that the career path suggested for Air Force space and missile operators prevents them from developing sufficient "depth," or expertise in any particular mission area. Instead, the goal is to have

---

192 Ibid.
193 Ibid.

95

brief assignments in as many mission areas as possible (touch all the bases or fill all the squares). This prevents sufficient maturation or "breadth" of expertise among officers gained by working in various jobs within a particular mission area. Next, he notes that "segregation" between research, acquisition, and operations communities—as well as with the National Reconnaissance Office—prevents necessary exchanges of ideas. Finally, McLaughlin points out that the shear size of the Missile Operations mission area, which accounts for two out of every three operational positions in Air Force Space Command are ICBM positions. He argues this makes it almost impossible to create officers with sufficient experience in specific space mission areas.[195] This is in stark contrast to the career specialization of aircrew members who typically spend several years operating one or at most a handful of aircraft.

To correct the shortfalls identified above and develop a cadre of true space experts that will become future space leaders, McLaughlin makes several recommendations. First, specific selection, training, qualification, and assignment criteria for all space-related positions would ensure the best-qualified personnel join the space career field. Second, combining current operations, research, and acquisition career fields would create space operators with greater depth in their profession. Third, creating separate career paths for each mission area to ensure leaders gain both depth and breadth within specialized areas will foster expertise. He stresses leadership involvement to ensure a sense of equality among the various mission areas. Next, he recommends recruiting space professionals with the types of technical degrees that fill the needs of the space career field, and giving Air Force Space Command centralized management of personnel

---

[194] Volume 28 of AFTTP 3-1 is the volume related to space systems.
[195] McLaughlin, 8-9.

in the space career field without having to go through the Air Force Personnel Center. Setting space professionals aside from the rest of the Air Force's personnel would be a significant because this would recognize the fundamental differences between the skills, mindsets, and operational specialties between airmen and space professionals. Finally, providing science-intensive training and education along the career path of a space professional would build better space leaders with the technical competency and intellectual diversity, they require.

Interestingly, McLaughlin does not address the training needs of the enlisted space professionals who comprise the majority of all space operators. They carry the brunt of the workload. Improving the recruiting criteria and training of enlisted space professionals must be included in any plan for accelerating America's spacepower excellence.

However space leaders are grown, the Air Force in particular must remember the unique global role of spacepower. The requirement for space support is not limited to Air Force operations in theaters where combat operations are underway. Instead, Air Force space assets and space professionals must simultaneously support *all* services in *all* theaters *all* of the time in *all* conditions of war and peace. This is fundamentally very different from the roles and missions of airmen. At the same time, space professionals must also take action to ensure relative control of the space medium to preserve friendly access while denying an adversary the ability to exploit space against supported nations.

While it is certainly essential for Air Force space professionals to be active partners in the planning and operations cells in Air Operations Centers around the globe, it is equally important for them to be present in the Army's corps-level planning centers and the Navy's fleet-level planning centers. While the Army and Navy have their own space personnel, they need equal access to the special expertise provided by Air Force space professionals as well. Similarly, Army and Navy space personnel would provide valuable expertise regarding their service's space operations in the Air and Space Operations Centers. As a rule, space professionals need to interact with warfighters in all other media. This will promote space integration across the board and yield new synergies in warfare.

Going to space is *still* hard and much of spacepower is *still* rocket science. Space is a place where several highly specialized systems perform a wide variety of missions. It is critical for space professionals, with their unique global mindset, to rapidly evolve into the leaders of military spacepower where they can set the spacepower vision for the future. When we recognize all of the dynamics that pertain to space operations, many of them captured in these propositions, we also realize that space professionals require career-long specialization.

#10 Weaponizing Space is Inevitable

*Space for peaceful purposes - what a bunch of *!#%*!.# bull*.!% that was!*

—Gen General Bernard Schriever,
USAF

It's politically sensitive, but it's going to happen. Some people don't what to hear this, and it sure isn't in vogue…but—absolutely—we're going to fight in space. We're going to fight from space and we're going to fight into space.

—Gen Joeseph W. Ashy
former USCINCSPACE

Weapons in space are inevitable, and the U.S. ought to review existing arms control obligations that get in the way of deploying a space-based deterrent.

— The Space Commission Report

Humanity has attempted to prevent or delay the proliferation of weapons for centuries, but history suggests that mankind is driven to develop new weapons.[196] For example, in 1139, at the Second Lateran Council, the Church banned the crossbow for being too lethal.[197] However, within a millennium, humans built nuclear weapons and used them in war. Competition is part of the human condition, and war is a natural expression of this condition. If this were not so, states would likely have forgone their military establishments and preparations for war a long time ago. We are not at the end of history—states still vie for power in anarchic international system and will compete in every medium of human endeavor. Former Secretary of the Air Force, Sheila E. Widnall states, "We have a lot of history that tells us that warfare migrates where it can—that nations engaged in a conflict do what they can, wherever they must."[198]

Space is already militarized by virtue of the force enhancements derived from current systems on orbit. Going one step further and weaponizing space is only a matter of time. Barry Watts believes weaponizing space may come about in one of two ways. First, there may be a dramatic trigger event, such as the use of nuclear weapons to attack orbital or terrestrial assets, which compels states to place weapons in space. Second, there may be a slippery slope wherein a series of small, seemingly innocuous

---

[196] The author wishes to sincerely thank Maj Scott Long for his contribution in writing this proposition.
[197] William H. McNeill, *The Pursuit of Power: Technology, Armed Force, and Society since A.D. 1000* (Chicago: The University of Chicago Press), 68.
[198] The Honorable Sheila E. Widnall, secretary of the Air Force, "The Space and Air Force of the Next Century," address to the National Security Forum, Maxwell AFB, Ala., 23 September 1997; on-line, Internet, 9 January 1998, available from http://www.af.mil/news/speech/current/The_Space_and_Air_Force_of_html.

developments in orbital capabilities over several years that would, in hindsight, be recognized as having crossed the boundary of weaponizing space.[199]

There is a growing national debate on the issue of weaponizing space, initially sparked by President Reagan's Strategic Defense Initiative and now reinvigorated by President George W. Bush's advocacy for a missile defense system. Although President Bush never mentioned placing weapons in space as part of his plan, his critics, such as Senate Majority Leader, the Honorable Thomas A. Daschle, claim this is implicit in his argument because space is the ideal place to station a small number of assets that can provide a global capability.[200] Pundits from both sides of the debate have forged the pros and cons of weaponizing space over the years. Frank Klotz describes the debate this way:

> On one side are those who argue that the United States needs to develop a military capability to protect its satellites from attack and to deny adversaries access to the benefits of satellite products and services. On the other side are those who contend that weapons should never be employed in space. They urge instead that arms control and other cooperative measures are the best means to protect American equities in space, as well as to prevent space from becoming an arena for armed conflict.[201]

Both sides of the debate have valid concerns. Proponents of weaponization claim the US will enhance its national power by weaponizing space. They are quick to point out that "there is no blanket prohibition in international law on placing or using weapons in

---

[199] Watts, 98.

[200] Senator Daschle called "placing weapons in space 'the single dumbest thing I've heard so far in this administration.'" As cited by James Dao, "Rumsfeld Plan Skirts Call for stationing Arms in Space," *New York Times*, 9 May 2001, on-line, Internet 12 May 2001, available from https://ca.dtic.mil/cgi-bin/ebird?doc_url=/May2001/e20010509rumsfeld.htm.

[201] Frank G. Klotz, *Space, Commerce, and National Security* (New York: Council on Foreign Relations Press, 1998), on-line, Internet 12 May 2001, available from http://www.foreignrelations.org/p/pubs/klotz.html.

space, applying force from space to Earth or conducting military operations in and through space."[202] Hays and Mueller describe this side of the debate like this:

> If the United States moves expeditiously to take advantage of its existing leadership in space technology and establish an unassailable dominance of orbital space, its position as the preeminent world power will be enhanced and perpetuated; if, on the other hand, it fails to seize the opportunity to establish unassailable superiority in space, its world leadership will be threatened by more visionary rivals…[H]e who controls space will control the world—or at least he who doesn't, won't—and, thus the more the United States invests in developing its spacepower, the more powerful and secure it will be.[203]

Conversely, advocates of preserving space as a peaceful sanctuary believe weaponizing space will reduce the overall power of the United States as an actor on the world stage. They are concerned about triggering security dilemmas that will lead to an arms race in space. According to Hays and Mueller:

> [Sanctuary realists] oppose space weaponization…because they believe it would reduce rather than enhance US power and security in particular. They argue that the United States, as the leading user of space, has far the most to lose if space systems become increasingly vulnerable to attack and that as the world's preeminent air and surface power, it has the least to gain from developing such weapons. Sanctuary realists also assert that if the United States takes the lead in developing space weapons, it will be easier for other states to follow suit, thanks to US technological trailblazing. Finally, they tend to be skeptical that the military utility of space weapons, both for power projection and to protect US space assets, will be as great as the weaponization proponents typically claim.[204]

Undoubtedly, there are numerous concerns over space-based weapons such as monetary costs, a questionable threat, lack of survivability, lack of political will, incompatibility with democratic values, problems with orbital dynamics and laser

---

[202] Space Commission Report, xvii.
[203] Hays and Mueller, 38-39.
[204] Hays and Mueller, 39.

physics, treaty infractions, and international opinion—just to name a few.[205] All these

concerns are serious and real. In fact, the solutions may not arrive for years, but they will

come to fruition sooner or later.

Regardless of which side of the argument is correct, the historical relationship

between man and his weapons provides insight into the probable future of space-based

weapons. Robert L. O'Connell suggests that human nature—not technology—is at the

root of weapons development.[206] Covetous motives required early man to develop new

ways to kill an opponent. Today's modern weapons are more lethal than the sticks and

stones of ancient days, but their purpose is still to gain an advantage over an adversary.

O'Connell suggests that humans will constantly develop new weapons as long as

their imagination discovers and exploits timeless and eternal scientific principles such as

quantum mechanics and relativistic physics, which may give them an advantage in

war.[207] Therefore, the very idea of weaponizing space becomes a driving force to do so,

like the idea of splitting and fusing the atom, made doing so inevitable. O'Connell states,

"Because so much of this is a function of the physical universe and the laws that govern

it, the process is, in a very real sense, beyond our control."[208]

Formulating new and creative ideas for weapons may be inevitable, but man always

has a choice whether to develop those weapons or not.[209] Dr. Colin S. Gray cautions that

the feasibility of deploying weapons in space does not mean that such weapons are

---

[205] See political concerns from Lt Colonel Bruce M. Deblois's article entitled, "Space Sanctuary: A Viable National Strategy" *Airpower Journal*, Winter 1998, n.p.; on-line, Internet, 4 December 2000, available from http://www.airpower.maxwell.af.mil./airchronicles/apj/apj98/win98/debloistxt.htm.
[206] Major theme from Robert L. O'Connell book entitled, *Of Arms and Men* (New York, NY: Oxford University Press) 1989.
[207] Ibid., 10.
[208] Ibid., 11.
[209] Ibid., 5.

strategically required—or prudent.[210] However, we are also warned that once someone begins development, the "technological imperative becomes nearly absolute," and that "Once the initial conditions are set, however, the logic of technology becomes nearly irresistible, at times sweeping us toward destinations never contemplated or desired."[211] This suggests the choice to weaponize space may be beyond rational decision-making. If this premise is correct, some actor may weaponize space as a poorly thought-out reaction to some unforeseen security dilemma, or may already be on the slippery slope towards weaponizing space as it seeks to protect its space systems—the two conditions that Watts believes will likely lead to the weaponization of space. The momentum is not likely to stop over the long run.

There is another sort of risk that makes weaponizing space much more likely. Without a world system of checks and balances that can ensure non-proliferation of space-based weapons, spacefaring states have the option of developing space-based weapons in a covert manner. Doing so might prevent an adversary from getting the upper hand. At the same time, secrecy would mitigate the risk of triggering a security dilemma and a subsequent arms race in space. An actor could place weapons on orbit claiming they were something else, or the actor could store the weapons on the ground in a launch-ready configuration. The later option simplifies the problem of maintaining secrecy, plus gives the opportunity to frequently inspect and upgrade the weapons. Unfortunately, this simple logic makes the secret development of weapons for space highly attractive—and much more likely.

---

[210] See the opening quote from Colin S. Gray and the discussion concerning the flawed strategy of the German V-2 program in Major William L. Spacy II's CADRE paper entitled, *Does the U.S. Need Space-Based Weapons?* (Maxwell AFB, Ala.: Air University Press, 1999), 9.
[211] O'Connell., 11.

It would be nice to assume that the notion of space-based weapons will just go away with the passage of time, but O'Connell suggests that *time* is the enemy. To fall behind and allow an adversary to gain the advantage would be detrimental to any actor's vital interests. Weaponizing space without careful strategic thought is not the answer, and a space-weaponization strategy is certainly not a "panacea" or a "single-point solution."[212]

Sad though it is, war is part of the human condition. Wherever humans go, they bring their cultural baggage with them, and weapons are part of the baggage. The utility of space is increasing in all areas of human activity and discourse. This may be a pessimistic view, but "we know from history that every medium—air, land and sea—has seen conflict. Reality indicates that space will be no different."[213] The political and military pragmatist must assume that over the long term the weaponization of space is inevitable.

---

[212] See practical considerations in Lt Col Bruce M. Deblois, "Space Sanctuary: A Viable National Strategy", 1998, n.p.; on-line, Internet 25 April 2001, available from http://132.60.50.46/airchronicles/apj/apj98/win98/deblois.html.
[213] Space Commission Report, 100.

Chapter 4

A Spacepower Theory

The Air Force has identified itself with the air weapon, and rooted itself in a commitment to technological superiority. The dark side of this commitment is that it becomes transformed into an end in itself when aircraft or systems, rather than missions, become the primary focus...Even though the Air Force is the lead military agency for space, space systems will be competing for aircraft roles and missions, posing difficult tradeoffs in budgets and force structure...[S]pace becomes a competing faction...

—Carl H. Builder

The Nature of Spacepower

The central question of this study is, "What is the nature of spacepower." There is no single answer to this question. Instead, the ten propositions regarding spacepower answer this question by revealing many of its characteristics. These are my findings:

1. Space is a Distinct Operational Medium. Physically space is very different from all Earthly media. More importantly, orbital operations are subject to unique sets of physical laws. This creates a wall of misunderstanding between space professionals and others who do not understand orbital mechanics. These physical differences heavily determine operational methodologies and special planning considerations for space activities. Most importantly, space was diplomatically set aside as a separate medium during the Eisenhower administration. The international community now recognizes entirely different legal standards in space. Every American administration and the DoD have reaffirmed the US belief that space is a separate and distinct operational medium. Only the US Air Force officially views air and space as a seamless operational medium.

2. The Essence of Spacepower is Global Access and Global Presence. The fundamental reasons for moving Earthly capabilities to orbit are to exploit the global nature of spacepower. Access to denied areas was the initial reason for putting up a satellite in order to conduct reconnaissance. This is still a compelling reason, however, the ability to conduct various missions globally with just a few assets is extremely compelling not only to the military and civil sectors, but especially to the commercial sector as well.

3. Spacepower is Composed of a State's Total Space Activity. Space activities have pushed beyond their intelligence and defense roots as states developed civil and commercial sectors as well. Going to space is difficult and it requires a substantial infrastructure to generate a space program. Spacefaring is most likely among wealthy nations that have a solid educational system that stresses the sciences, natural resources, stable political environment, and a strong will to commit to a space program over the long term.

4. Spacepower Must Be Centrally Controlled By a Space Professional. Spacepower is different from other forms of military power because its missions are global in scale. Because spacepower assets are global in nature, it would be wrong to manage them from a theater perspective, as is the case with most terrestrial forces. Doing so would handicap spacepower in the same way airpower was handicapped at the outbreak of the Second World War when it was broken into penny packets under army control. A space professional with a global mindset must centrally control spacepower in order to balance scarce resources across theaters. At the same time, space professionals must take charge of the battle for space control, not leaving this responsibility to other commanders with different priorities and concerns.

5. Spacepower is a Coercive Force. The mere presence of spacepower assets such as reconnaissance and surveillance satellites has already and will increasingly influence the activities of actors wishing to conceal certain activities. This flows from the deterrent potential of collection assets that have long been used as national technical means of treaty verification. It is quite likely that some actors are deterred from certain courses of action in the presence of spysats. Increasingly spacepower assets are integrating into the sensor-to-shooter loop of active combat operations. This, plus the inevitable emergence of weapons on orbit, signals the expansion of spacepower's coercive force into the role of compellence in addition to deterrence.

6. Commercial Space Assets Make All Actors Space Powers. The advent of commercial vendors selling military-related space products has created a new form of mercenary. The types of asymmetric advantages the superpowers once enjoyed because of their space prowess is quickly eroding because anyone who is able to pay the price can receive certain kinds of space support. Military and law enforcement planners must take into account the potential for any opponent to exploit these commercial services.

7. Spacepower Assets Form a National Center of Gravity. More and more segments of society are turning to space-based assets for services. This makes the relatively few satellites on orbit very lucrative targets for an adversary who has the will and means to strike them. While satellite access is seldom a single point of failure, losing access to the vital information collected and carried by them will increase the fog, friction, and cost of operations. In certain circumstances, this may turn the tide against spacefaring states, such as the US.

8. Space Control is Not Optional. The increasing reliance on spacepower assets by the government, intelligence, military, and business segments of society make it essential to secure access to satellite services. At the same time, it is equally important to deny an adversary access to their space systems to increase their fog, friction, and cost. Adversaries will likely compete for relative control of the space medium, therefore measures must be taken to secure national interests in space.

9. Space Professionals Require Career-Long Specialization. Going to space is still hard. Despite more than forty years of spacefaring experience, we still face numerous technical challenges. Moreover, space operations are so different from any form of terrestrial operations that developing space experts requires highly specialized and recurring education, as well as careful career management.

10. Weaponizing Space is Inevitable. Wherever mankind goes, weapons follow. There are some rock solid reasons for not weaponizing space, but they fail to take into account the technological imperative that often drives human behavior in ways frequently beyond rational thought. When weapons will be placed in space is anybody's guess, but the political and military pragmatist must assume that someone will put weapons in space and plan accordingly.

The Hypothesis is Rejected

This study *rejects* the hypothesis that spacepower is merely a continuation or extension of airpower, as advocates of the aerospace integration school claim. Spacepower *is* an independent form of power that directly affects all other national instruments of power, and increasingly affects the daily lives of ordinary humans. As a form of military power, spacepower is of growing importance because it forms a global informational infrastructure that the armed forces of advanced nations increasingly rely upon. In the future, spacepower will likely include counterspace weapons and systems designed to attack terrestrial targets.

Advocates of the aerospace integration school may fairly criticize this study as stacking the deck against their case simply because using the term "spacepower" concedes an independence from other types of power. This is a valid form of criticism. The logic behind this criticism is just as valid now as it was at the court martial of Billy Mitchell who brazenly argued for something called "airpower."

Aerospace integrationists frequently argue that spacepower is in no way different from airpower because it delivers similar products to users, as if aircraft can do what spacecraft can do. This is simply not the case. Aircraft cannot survey more than 80 percent of the Earth with only three aircraft, nor can they freely pass over denied airspace. Spacecraft can. Only a handful of satellites can provide persistent capabilities on a global scale. The Global Positioning System (GPS), for example, employs only 24 satellites in its nominal constellation, yet it is described as "the first global utility." Aircraft simply do not do the types of missions on a global scale that spacecraft perform very efficiently. While it is true that some satellites perform similar missions as aircraft, such as photoreconnaissance, reconnaissance aircraft loiter over theater-specific areas of interest, while reconnaissance satellites transit the globe in a matter of minutes taking photographs in virtually every theater along their route. Not only are air and space assets employed differently, but the fundamental differences between airpower and spacepower create unique operational mindsets among operators. Airmen posses a theater level perspective, whereas space professionals posses a global mindset.

A predictable response from the aviation community is repeated attempts to make an aircraft that can do what a spacecraft can do. There have been many attempts to produce a spaceplane, but none have become operational, in part because of technical problems, but mostly because no one could ever explain why an enormous sum of money should be spent to make an aircraft do what a satellite can do (other than serve as a reusable spacelift and recovery vehicle). Currently there are arguments that America needs a combat-capable spaceplane in order to deliver ordnance more rapidly than aircraft, and to do so without requiring forward bases in a combat zone. These are compelling ends, but justifying the means will likely be as difficult today as it was at the height of the Cold War.

Lessons Learned

1. Military Space Doctrine Is Highly Controversial.

The highly politicized nature of space doctrine stems from its clandestine roots in the spy vs. spy era of the Cold War.[214] There is undeniable tension between those who believe US power interests are best served by preserving space as a peaceful sanctuary and those who believe US power interests demand the rapid development and fielding of weapons in space. The recent international debate regarding weaponizing space as part of President Bush's missile defense program indicates the global sensitivity to American military space strategies. This indicates that weaponizing space will bring about security

110

dilemmas, unlike introducing new weapons into the traditional media of air, land, and sea.

2. Spacepower Decision-Making Is Fragmented and Highly Bureaucratic.

There is no single advocate for spacepower. USCINCSPACE shares his authority with numerous agencies—each having a say in space strategy. Ultimately, spacepower policymaking rests with the NCA, but the military space program is a house divided with many voices advocating different desires. The Space Commission recommended realignment of the many agencies in the military space community to help consolidate authority into a smaller group of decision-makers.

3. The US Air Force Is Out Of Step with the National and Defense Space Policies.

The aerospace integration effort is at odds with national and DoD policies that identify space as a separate medium coequal with air, land, and sea. Despite more than forty years of pushing the term aerospace, nobody is using it, at least not outside the Air Force where the issue of space being a separate medium is already settled. This hurts the credibility of airmen who advocate for spacepower and does spacepower a grave disservice since these advocates typically lack any degree of space expertise. The quote by Carl Builder that begins this chapter is partly correct, the Air Force is institutionally thrashing within itself over a budgetary and philosophical rivalry with the nation's spacepower. On a programmatic level, space and air systems do compete for the limited resources in the Air Force's budget and space systems do not always fare well in this competition. On a doctrinal level, however, Builder is wrong, like so many airmen, to

---

[214] It is interesting to speculate how spacepower might have evolved if it came of age during a total war such as World War I or II. If the airpower analogy applies, then it is probably safe to presume that space

111

assume that airpower and spacepower compete with each other for roles and missions. This is simply not the case. Airpower continues to provide theater-focused forces, whereas spacepower provides globally focused forces. The two compliment each other as joint partners, along with land and sea forces.

4. American Grand Strategy Hinges On the Assumption Of Space Superiority.

The successful integration of space forces with terrestrial activities makes all of our instruments of power increasingly reliant on space support. Without space control capabilities to defend friendly access to space while denying the same to an adversary, we rest our case on hope. The Space Commission recommends in unambiguous language that the US must develop and field space control capabilities to safeguard national interests.

5. The Space Commission May Be the Deus Ex Machina.

One of the basic findings of the Space Commission is that "The US Government—in particular, the Department of Defense and the Intelligence Community—is not yet arranged or focused to meet the national security space needs of the 21st century."[215] In one sense, the Commissioners concluded that the military space community is broken. The Space Commission made several recommendations that may fix the deficiencies they identified. While it did not specifically address the Air Force's aerospace integration policy (in fact, the term was conspicuously absent), it does recommend moving space experts into leadership positions within the space community from which they will advocate for spacepower. The reform-minded nature of the Bush Administration makes

would now be fully weaponized and the global military power structure might be radically different.

such reforms likely, especially since the chairman of the Space Commission is now serving as the Secretary of Defense. The specific findings and recommendations of the Space Commission appear in Appendix E of this study.

A Spacepower Theory

The introduction of this study posited that the ten propositions regarding spacepower might serve as a foundation to assist in the creation of a spacepower theory. Such a theory would serve political and military practitioners by providing a framework to assess spacepower issues and guide their related decision-making. Presented here is an attempt to synthesize the ten propositions and lessons learned during this study into the rough sketch of a spacepower theory.

Spacepower is a tool of statecraft and warfare, therefore any theory of spacepower must be firmly rooted in broader theories of statecraft and warfare. Instead of rehashing what the time-honored masters have already done so brilliantly, suffice it to say here that students of spacepower should read Sun Tzu's *The Art of War*.[216] This book crisply describes the nature of statecraft and its nexus with warfare in a world where states are in constant competition with one another. Next, one should read Carl von Clausewitz' treatise *On War*.[217] This book, although difficult to read and easy to misinterpret, captures the central premise of war when he states, "war is nothing but the continuation of policy with other means," that "the political object is the goal, war is the means of reaching it, and the means can never be considered in isolation from their

---

[215] *Report of the Commission to Assess United States National Security Space Management and Organization* (Washington D.C., January 11, 2001) ix.

[216] There are several translations and editorialized versions available. The author recommends Sun Tzu, *The Art of War*, trans. Ralph D. Sawyer (New York: Barns and Noble Books, 1994).

purpose."[218]   With these intellectual underpinnings, and an appreciation of the propositions regarding spacepower, students of spacepower are ready for a theoretical discussion.

The purpose of military spacepower is to provide global capabilities to assist in achieving political and military objectives. It is an independent form of power that can be used alone or in concert with other forms of power to achieve desired ends. Space is a place where humans station systems that help resolve problems. It begins above the Earth's surface at the lowest altitude where a satellite can sustain a circular orbit (approximately 93 miles) and extends outward to infinity. Eventually, humanity may extend its interests beyond near-Earth space. Military spacepower will likely protect those interests. Someday, far into the future, populations and their political entities may migrate into space as well. For now, however, humans live on the surface of the Earth, and contemporary spacepower in this context refers to the struggles occurring there, but this will evolve over time.

The reason for going to near-Earth space is to gain access to regions of the Earth where terrestrial forces either cannot go, or cannot loiter as economically as some satellites. A relatively small number of similar satellites spread out in orbital space can survey the entire Earth's surface, which gives space-based constellations the ability to perform missions on a global scale. States perform many missions in space. In the opening years of the 21st century, these missions are primarily informational, i.e., providing command and control, communications, and computer support (C4), and intelligence, surveillance, and reconnaissance (ISR) support to terrestrial forces. Air,

---

[217] Carl von Clausewitz, *On War*, ed. and trans. Michael Howard and Peter Paret (Princeton, N.J.: Princeton University Press, 1976).

land, and sea forces also perform missions like these, but only space systems (and some terrestrial communications networks) perform them continuously on a global scale. These space networks create a global informational infrastructure that links together expeditionary forces deployed anywhere in the world and connects these forces with their homeland leadership. Terrestrial forces are predominantly concerned with performing these missions inside relatively small theaters of operation. In the future, humans will not only employ space-based weapons to gain control of space, but humans will also employ them against targets on the surface of the Earth, at sea, and in the air. With sustained national commitment to technological advancement and investments of time, talent, and treasure, space will continue to provide an excellent vantage point from which to observe, support, and influence human events, but space systems require a vigorous defense.

Space control is job one. The first and most enduring mission of space forces is to gain relative space control over adversaries when hostilities occur. This means providing continuous situational awareness about what is happening in space (space surveillance), ensuring friendly access to space and the support provided from space systems, while denying the same to any adversary. Space control has both a defensive and an offensive component.

Defensive space control efforts must ensure friendly forces and their political leaders can continue to exploit space for three reasons. The first reason is to support theater operations where combat operations are underway. The second reason is to continue observing activities in all other theaters to assess other potential threats requiring diplomatic or military intervention. The third reason is to sustain continuous global services from space, such as communications and Global Positioning System data upon

---

[218] Ibid., 69 and 87, respectively.

which users in all theaters are increasingly reliant. At the same time, commercial assets in space require some degree of defensive protection. Ideally, all satellites should be hardened from attack, but commercial investors are reluctant to spend the money to protect their satellites. This places a burden on defense planners to provide some protection to commercial systems that are important to the business interests of domestic and allied economies.

An adversary's counterspace weapons may be able to attrit friendly space systems very rapidly, therefore it is imperative for space powers to acquire the ability to find, fix, track, target, and destroy an adversaries counterspace weapons very quickly. Such systems may reside on land, at sea, in the air, or in space. It is equally imperative for space powers to acquire the ability to repair or replace lost satellite services on orbit. The goal is to rapidly restore space support before it affects political and combat operations. Activating on-orbit spares, leasing commercial satellite services, launching new satellites to replace those lost through attrition, or gaining access to an ally's satellite services may do this. It is also essential for space powers to acquire the ability to repair or replace lost satellite ground control systems. Methods for doing this may include transferring ground control responsibility to another location (fixed or mobile), leasing commercial support, or obtaining ground support from an allied state.

On the offensive side, space control does not need to be total in order to be effective. For example, the enemy may have satellites that do not especially affect their warfighting ability or influence the outcome of a conflict. The situation and strategy will dictate the degree of offensive space control that is required. Factors to consider will be the time and place where space control must be gained, how rapidly it is needed, the number of satellites or ground control targets requiring negation, how long space control must be sustained, and the desired level of negation, e.g. destruction, degradation, denial, disruption, etc.

It is important to remember that an adversary's satellites are global assets. It may be politically untenable to permanently damage an adversary's satellite for a number of reasons. For example, while an imagery satellite may threaten to disclose friendly troop movements in one region, that same satellite might perform treaty verification on the opposite side of the globe or other missions that there is a friendly interest in preserving. In many scenarios, offensive space control might best be limited to very localized and temporary effects.

The best way to deny an adversary access to space is to destroy his space launch facilities, but we must also be aware that the adversary may contract his spacelift with other countries where he may have satellites in storage. The best way to deny space support to an adversary is to directly negate the satellites he uses. While some satellite systems may be particularly susceptible to the destruction of their ground stations, this may have only limited effect on other satellite systems that may degrade gracefully in the absence of ground control. It is also likely that an adversary will employ mobile ground stations for tactically important space systems that require frequent ground control. This not only makes targeting ground stations more difficult, but it highlights the need to negate an adversary's satellites on orbit. It is also possible to attack the users of space support by jamming or spoofing their receivers. This has the benefit of localized and temporary effects. In many scenarios, it is likely that a combination of attacks on all three segments of a space system (ground stations, satellites on orbit, and user equipment), as well as their linkages, will be required to achieve the desired effect.

Space control efforts will be complicated if an adversary is using third-party launch facilities, satellites, or ground control systems provided by commercial vendors, international consortia, or an ally. Diplomatic efforts will likely be required to eliminate third-party support to adversaries, but if the political will exists, friendly forces must be ready to expand the conflict by striking wherever adversaries receive space support. If diplomatic efforts fail and policy does not allow expansion of the conflict to strike third-party targets, then the adversary has a sanctuary he will likely exploit.

Situational awareness is paramount. Situational awareness has always been critical in diplomacy and warfare, but in the new era of precision targeting, situational awareness must be equally precise—a bomb is only as accurate as the coordinates used by the planner, the warfighter, and the munition itself. Precision targeting is well understood, but the need for precision ISR is not.

Multitudes of ISR sensors in all media characterize the modern battlespace. Some collect signal intelligence while others collect photoreconnaissance data. Still others collect radar information. These sensors and their operators not only attempt to identify targets, but also try to determine each target's precise coordinates. The ability of different sensors to determine the precise coordinates of targets varies, but in general, airborne sensors are much better at this than space-based sensors for several reasons. First, space systems are typically much further away from the targets than aircraft. Second, satellites in the lowest orbits are moving very fast in relation to targets and have relatively short dwell times on targets compared to aircraft, and satellites in higher orbits are much more distant and are generally less able to refine target coordinates as precisely. Third, satellite sensors degrade over time and presently there is no effort underway to perform physical maintenance on them to keep them in prime condition. Finally, given the relatively few ISR satellites in low-Earth orbit, continuous coverage of areas of interest from space with the most precise space-based sensors is currently impossible. In sum, aircraft have several distinct advantages over spacecraft in regards to *theater* ISR collection, but space-derived ISR data is critical to operations.

Space-derived ISR data is critical to diplomatic and military operations because it provides a "first look" at the battlespace and assists planners in finding and coarsely geolocating many targets before terrestrial forces move into the region. As a rule of thumb, today's space-derived ISR is useful in finding 80 percent of the targets and is able to determine their location to roughly 80 percent of the accuracy required to conduct precision strikes. In some cases, space systems do better than 80 percent in finding and fixing targets, and in other cases, they do worse. What is important is the tremendous advantage space systems provide politicians and commanders by giving them a good quality first look into the situation they face. With this information, they are able to make decisions about how best to employ their limited terrestrial ISR assets (aircraft, ships, submarines, reconnaissance ground forces, etc.) more efficiently to refine the ISR picture to the quality they desire for the operations they are considering. In some cases, the first look from space may suffice, but usually terrestrial ISR assets are required. During combat operations, space-based ISR sensors continue to provide data, filling gaps in coverage by theater assets. Space-based ISR sensors also frequently cue terrestrially

119

based sensors, as was the case during the Gulf War with missile warning satellites cuing Patriot batteries to intercept Iraq's inbound Scud missiles.

Perhaps most important of all, day in and day out, during war and peace, spacepower provides the 80 percent first look on a *global* scale. It allows analysts to watch the world and report tip-offs, warnings, and indications that give political and military leaders the freedom to employ their terrestrial forces more expeditiously, and with greater confidence that another threat is not more pressing. Spacepower literally watches the backs of terrestrial forces to make sure no threat is sneaking up behind them. This allows greater concentration of terrestrial forces in theaters of combat operations because space-based ISR assets are sufficient to act as a kind of global sentry. This sort of mission is ideally suited to space systems because they have unimpeded access around the globe and relatively few assets are required to sustain ISR missions on a global scale.

Much more is possible. By increasing the numbers of low-Earth orbiting sensors, continuously improving the quality of the sensors, and developing the means to service and repair them (either on orbit or by recovery and re-launch), the 80 percent rule of thumb will creep closer toward the 100 percent solution, despite the warfighter's demand for ever increasing precision. As space systems becomes more capable, is it likely that they will replace terrestrial forms of ISR collection? No. Aerial reconnaissance did not eliminate the need for land and sea forces to conduct reconnaissance of their own. There is no reason to believe that space-based reconnaissance will replace any other form of reconnaissance either.

Spacepower does not usurp missions from other forces. Spacepower assets give a state new core competencies for its military order of battle. The ability to do anything

120

continuously on a global scale is a new contribution to warfare made possible by spacepower. The various C4ISR capabilities, including weather observation, missile warning, and navigation and timing broadcasts give American forces a distinct asymmetric informational advantage over adversaries in the opening days of the 21st century. This advantage will evaporate over time as other actors on the world stage develop, lease, or borrow similar capabilities.

Despite having to share much of their budget with the US Air Force, American space forces do not compete with terrestrial forces for roles and missions. Airpower, landpower, seapower, and now spacepower (and possibly information power) bring different capabilities to modern warfare. The US trains its military members in highly specialized ways with the objective of being able to dominate operations within their respective media. Operations in each media require centralized control by practitioners of that form of power, in close coordination with the other warfighters, to ensure the optimum management of resources.

A great fallacy resulting from the aerospace integration mindset is the oft-cited statement that "airpower missions will migrate to space when it becomes reasonable to do so." This presumes that theater commanders are willing to trade highly flexible organic airpower assets for less flexible (and often less capable) space systems that another commander would likely manage as global assets. Economic considerations may force such a compromise, but a more prudent approach is to develop robust space capabilities in addition to airpower, landpower, and seapower assets. Remember, the difference between space systems and terrestrial systems is that space systems provide global access and global presence during both war and peace. Terrestrial systems should be developed as organic theater assets to fill gaps in space coverage and provide more flexible and precise ISR data and strike capabilities.

When space forces eventually obtain systems that can create physical effects at any location on the surface of the Earth, e.g. conventional bombing or "rods from God," this will not replace the standing requirement for aircraft to be able to do the same thing. Space operations are expensive and economic considerations may require air delivery of munitions. Exceptions include times when cost is not a consideration, such as combat in areas where aircraft are denied access, when aircraft cannot respond to a time-critical situation as quickly as spacecraft, when only a specialized weapon delivered from space will have the desired probability of killing a target, and when surprise is of the utmost importance.

There is unquestionably some overlap between the capabilities of spacepower and other forms of power, but this is a source of strength, not waste. Just as the triad of bombers, submarines, and missiles during the Cold War prevented an adversary from gaining a significant advantage should they successfully counter one of the legs, today's redundancy prevents an adversary from gaining a significant advantage should they successfully counter space-based systems or other terrestrial forces. There will be some adjustments in force structures as space capabilities become more robust, but *no* mission in *any* service should *ever* move *entirely* to space. Under *no* circumstances should *all* of the eggs *ever* be placed in the space basket. Instead, there should be an integrated combined arms approach.

The combined arms approach. During time of peace, spacepower assets monitor the globe, helping to identify and characterize potential threats. When a threat emerges, political and military leaders may opt to send terrestrially based ISR sensors into the area of interest to get a closer look. Should hostilities break out, space forces will gain whatever space control is required and will contribute whatever they can to help friendly forces in theater in terms of ISR and strike capabilities, but they still must watch the rest of the world, in every other theater, looking for tip-offs, warnings, and indications of other threats.

Force application from space will take many different forms, but it seems likely that space-based weapons will fill specific niches, ideal for a handful of missions during certain phases of operations. No claim is made that spacepower by itself can be decisive in general conventional warfare, but in certain circumstances, it may help set the conditions for victory by friendly forces. Conversely, if spacepower forces are defeated, this may turn the tide of the war against friendly forces and contribute to defeat. There may be certain forms of limited warfare where the information gleaned from space or strikes from space may achieve the political and military aims of an operation. If this defines decision in the battlespace, then so be it.

Unfinished Business

This study identified and argued the case for ten propositions regarding spacepower. Doing so revealed that air and space, indeed, airpower and spacepower are different. With a greater understanding of the nature of spacepower, it became possible to construct a spacepower theory. This theory presented above supports the timeless principles of statecraft and warfare, but it is also complementary—not competitive—with other forms of power. There is the unfinished business of building spacepower for the nation that matches the vision laid out in this theory.

It is beyond the scope of this study to recommend who should build spacepower for the nation. There are many possibilities. Should it be the Air Force? Should it be a Space Corps within the Air Force filled with dedicated space professionals whose only focus is spacepower? Should it be a joint effort under US Space Command? Should it be the responsibility of an independent Space Force? The nation should address these questions in light of the ten propositions above. Readers of this study should be better able to develop an informed opinion and make spacepower decisions accordingly.

Appendix A

Oberg's 12 Truths and Beliefs About Spacepower

James E. Oberg completed the book *Space Power Theory*, based on a draft from Dr Brian R. Sullivan. Oberg worked for a NASA contractor at the Johnson Space Center as a space engineer from 1975 to 1997. He is an author of several space-related books and magazine articles. He is also a consultant to news organizations, commercial corporations, the US military and the Congress on space-related subjects.[219]

The book was commissioned by General Howell M. Estes III, who served as USCINCSPACE from 1996 to 1998. The General wanted a book that would stimulate the spacepower debate in the same way the works of Douhet, Trenchard, and Mitchell did for airpower.[220] As a result, the book touches a wide range of topics such as contemplating the importance of spacepower, legal issues, the nature of the space environment, orbitology, current and future capabilities, organizational issues, etc.

Here are Oberg's twelve "Truths and Beliefs" about spacepower[221]

1. The primary attribute of current space systems lies in their extensive view of the Earth.
2. A corollary to this attribute is that a space vehicle is in sight of vast areas of the Earth's surface.
3. Space exists as a distinct medium.
4. Space power, alone, is insufficient to control the outcome of terrestrial conflict or ensure the attainment of terrestrial political objectives.
5. Space power has developed, for the most part, without human presence in space, making it unique among other forms of national power.
6. Technology competence is required to become a space power, and conversely, technological benefits are derived from being a space power.

---

[219] James E. Oberg, *Space Power Theory* (Washington, D.C.: U.S. Government Printing Office, March 1999), back cover.
[220] Ibid.,vi.
[221] Ibd., 124-131.

7. As with Earthbound media, the weaponization of space is inevitable, though the manner and timings are not at all predictable.
8. At some time in the future, the physical presence of humans in space will be necessary to provide greater situational awareness.
9. Situational awareness in space is a key to successful application of space power.
10. Control of space is the linchpin upon which a nation's space power depends.
11. Scientific research and exploration pays off.
12. There will be wild cards.

Appendix B

Gray's 8 "Clausewitzian" Ideas About Space Power

In his book, *Modern Strategy*, author Colin S. Gray promotes his central thesis that "there is a unity to all strategic experience: nothing essential changes in the nature and function (or purpose)—in sharp contrast to the character—of strategy and war."[222] A professor of International Politics and Director of the Centre for Security Studies at the University of Hull, Gray has researched and published numerous books and articles on strategy. His recent interest lies in the impact of evolving technology upon modern warfare. In particular, he has spent much effort contemplating the role of spacepower and the need for spacepower theory upon which to base the particular aspects of spacepower strategy. Particulars aside, he contends that the roots of all strategy can be found in the logic laid down by Clausewitz in the timeless classic, *On War*. As such, Gray believes that the logical underpinnings for spacepower theory necessarily must have their roots planted firmly in the Clausewitzian tradition.

Here are Gray's "Clausewitzian ideas" about spacepower:[223]

1. War has a grammar, but not a policy logic of its own.[224] War in space has its own distinctive characteristics that policy must know and respect, but such war has meaning only for the purposes of policy.
2. Countries have 'centres of gravity' key to their functioning.[225] A country's or coalition's ability to wage war successfully can be negated if those centres of gravity are menaced, damaged, or taken. Space forces can greatly enhance the

---

[222] Colin S. Gray, *Modern Strategy* (Oxford, Oxford University Press, 1999), 256-257.
[223] Ibid., ix.
[224] Carl von Clausewitz, *On War*, ed. and trans. Michael Howard and Peter Paret (Princeton N.J.: Princeton University Press, 1984), 605.
[225] Ibid., 595-7.

ability of other kinds of military power to locate, threaten, harass, and destroy such centres.

3. War is the realm of chance, uncertainty, and friction; the fog of war blinds the commander.[226] Spacepower assaults some of the friction that impairs terrestrial military performance, but is itself subject to the workings of friction.

4. War is a unity.[227] Spacepower is an essential team player, probably due to become the team player who adds the greatest value for lethality in combat in the twenty-first century.

5. Policy-makers and military commanders need to understand what the military instrument can accomplish under particular conditions.[228] The emergence of spacepower adds to the burden of comprehension by military professional and civilian layperson alike.

6. As the Just War tradition maintains, there needs to be a unity of character and intensity of political propose with the scale and kinds of military means: the principle of proportionality.[229] Contemplation of the military implications of a maturing spacepower has to accommodate appreciation of the value to policy of an unprecedentedly discriminate military instrument, without being captured by techno-military fantasies.

7. Success in battle flows from the achievement of overwhelming strength at the 'decisive point.'[230] This maxim is as sound for space operations as it is for other kinds of military activity.

8. Defense is the stronger form of waging war (on land).[231] In space, defence is probably the stronger form of waging war in high- and medium-Earth orbit (HEO and MEO), but probably not in low-Earth orbit (LEO). There is some safety in sheer distance (equal to time, provided speed-of-light-directed energy weapons are not relevant).[232]

With its Clausewitzian roots framed, Gray argues that the elements of spacepower theory can be assembled in piecemeal fashion from theories of airpower and seapower because the logic of warfare in one medium applies to warfare in all other media.[233] His list of Clausewitzian ideas implies that space is a distinct medium and potentially a center

---

[226] Ibid., 119-21.

[227] Ibid., 607.

[228] Ibid.

[229] Ibid., 88, 579.

[230] Ibid., 204.

[231] Ibid., 359.

[232] Note that "HEO" is typically used in America as an acronym for "Highly Elliptical Orbit." Gray, however, is British and uses the acronym "HEO" to mean "High Earth Orbit." HEO, in the sense used by Gray, is an orbit that extends beyond 35,000 km in altitude; MEO extends from 800 to 35,000 km; LEO extends from 150 to 800 km.

[233] Colin S. Gray, *Modern Strategy* (Oxford University Press, Oxford, 1999) 257.

of gravity. As such, the employment of space assets requires special policy and military considerations in order to exploit it properly—just like the other media.

Appendix C

Gray's 7 Most Vital Assumptions about Space Power

Having established the Clausewitzian roots as the framework for a spacepower theory, Gray points out that air, land, sea, and space have much in common. This must be so if he is to prove his thesis that there is unity to all strategic experience.

Here are Gray's seven most vital assumptions about space power:[234]

1. In all strategic essentials, spacepower is akin to landpower, seapower, and airpower.
2. The strategic history of spacepower is likely to follow the pattern already traced by seapower and airpower.
3. Geographically, space is distinctive, but then so is the land, the sea, the air, and even cyberspace.
4. People have only one natural environment, the land. To function in any other geography, they require technological support. The vacuum of space admittedly is exceptionally hostile to human life, but it does not differ basically in character from the sea and the air: all these geographies can tolerate human presence only when that presence is supported by machines.
5. Because people live only on the land, and belong to security communities organized politically with territorial domains, military behaviour, no matter what its tactical form, ultimately can have strategic meaning only for the course of events on land. It follows that seapower, airpower, and now spacepower function strategically as enabling factors. The outcome of a war may be decided by action at sea, in the air, or in space, but the war must be concluded on land and usually with reference to the land.
6. The logic of strategy is geographically universal and temporally eternal. Different strategic cultures may 'do it their way', but only if that way is consistent with the laws of physics, *inter alia* (willpower is only hot air, if the engineering is unsound).

---

[234] Ibid., 258-259.

130

7. The unique geography of space must find expression in unique technology, operations, and tactics. That unique geography does not, however, point the way to some unique logic of strategy, let alone a unique irrelevance of strategy.

Appendix D

Mantz's 10 Axioms of Space Combat Power

In May, 1995, Lt Col Michael R. Mantz published an Airpower Research Institute sponsored research report titled, "The New Sword: A Theory of Space Combat Power. The purpose of the report was to present an "unconstrained and comprehensive" theory of space combat power.[235] The report is essentially an elaborate think-piece drawing on Mantz's extensive academic and practical experience. He graduated from the Air Force Academy in 1976 with a bachelor of science degree in electrical engineering. Mantz completed his master of science degree in aeronautics and astronautics from the Massachusetts Institute of Technology in 1981 before his selection and subsequent training as a space shuttle payload specialist (astronaut) in 1982.

In an appendix to his work (pages 73-82), Mantz included what he called, "Axioms of Space Combat Power," a list of ten statements very similar to propositions. He included them in his work because "any theory uses axioms as building blocks."[236] Essentially, he framed propositions that dealt primarily with the concept of space combat, vice spacepower as a whole.

---

[235] Michael R. Mantz, Lt Col., "The New Sword: A Theory of Space Combat Power" (Air University Press, Maxwell AFB, May 1995) xi.
[236] Ibid., 74.

Here is Mantz's list of axioms:

1. Space strike systems can be employed decisively by striking Earth forces, both independently and jointly
2. Space strike systems can be employed decisively in war when the enemy's essential means for waging war (industry, transportation, and communications) are vulnerable to attack from space.
3. Space strike systems can be employed decisively by striking at the decision-making structure (leadership and command and control) of the enemy.
4. Space strike systems can deter hostile actions by holding forces, decision-making (leadership and command and control), and infrastructure (industry, transportation, and communications) at risk.
5. Space denial systems can be employed decisively by denying enemy access to space-derived data.
6. Space denial systems can be employed decisively by physically denying enemy access to space.
7. Space protection systems can be employed to assure friendly access and use of space.
8. Total space control (the combination of space denial, space protection, and passive space defense measures) is neither achievable nor necessary.
9. Space combat power must be centrally and independently controlled.
10. Space power is not intrinsically linked to airpower. [237]

Mantz goes on to provide an excellent discussion that links his axioms to his theory of space combat. It is important to note that Mantz is of the opinion that space is a distinct medium of operations and spacepower has an independent mission separate from its role of supporting or augmenting surface forces. Interestingly, he also states his belief that space control is neither achievable nor necessary.

Mantz remains on active duty in the US Air Force and continues to serve as a space professional.

---

[237] Ibid.

Appendix E

Findings and Recommendations of the Commission to Assess United States National Security Space Management and Organization (Space Commission Report)

The Commission has unanimously concluded that organizational and management changes are needed for the following reasons.

First, the present extent of U.S. dependence on space, the rapid pace at which this dependence is increasing and the vulnerabilities it creates, all demand that U.S. national security space interests be recognized as a top national security priority. The only way they will receive this priority is through specific guidance and direction from the very highest government levels. Only the President has the authority, first, to set forth the national space policy, and then to provide the guidance and direction to Senior officials, that together are needed to ensure that the United States remains the world's leading space-faring nation. Only Presidential leadership can ensure the cooperation needed from all space sectors—commercial, civil, defense and intelligence.

Second, the U.S. Government—in particular, the Department of Defense and the Intelligence Community—is not yet arranged or focused to meet the national security space needs of the 21st century. Our growing dependence on space, our vulnerabilities in space and the burgeoning opportunities from space are simply not reflected in the present institutional arrangements. After examining a variety of organizational approaches, the Commission concluded that a number of disparate space activities should promptly be merged, chains of command adjusted, lines of communication opened and policies modified to achieve greater responsibility and accountability. Only then can the necessary trade-offs be made, the appropriate priorities be established and the opportunities for improving U.S. military and intelligence capabilities be realized. Only with senior-level leadership, when properly managed and with the right priorities will U.S. space programs both deserve and attract the funding that is required.

Third, U.S. national security space programs are vital to peace and stability, and the two officials primarily responsible and accountable for those programs are the Secretary of Defense and the Director of Central Intelligence. Their relationship is critical to the development and deployment of the space capabilities needed to support the President in war, in crisis and also in peace. They must work closely and effectively together, in partnership, both to set and maintain the course for national security space programs and to resolve the differences that arise between their respective bureaucracies. Only if they do so will the armed forces, the Intelligence Community and the National Command Authorities have the information they need to pursue our deterrence and defense objectives successfully in this complex, changing and still dangerous world.

Fourth, we know from history that every medium—air, land and sea—has seen conflict. Reality indicates that space will be no different. Given this virtual certainty, the U.S. must develop the means both to deter and to defend against hostile acts in and from space. This will require superior space capabilities. Thus far, the broad outline of U.S. national space policy is sound, but the U.S. has not yet taken the steps necessary to develop the needed capabilities and to maintain and ensure continuing superiority.

Finally, investment in science and technology resources—not just facilities, but people—is essential if the U.S. is to remain the world's leading space-faring nation. The U.S. Government needs to play an active, deliberate role in expanding and deepening the pool of military and civilian talent in science, engineering and systems operations that the nation will need. The government also needs to sustain its investment in enabling and breakthrough technologies in order to maintain its leadership in space.[238]

Following are the Commission's unanimous recommendations.

1. Presidential Leadership

The United States has a vital national interest in space. National security space should be high among the nation's priorities. It deserves the attention of the national leadership, from the President down.

---

[238] *Report of the Commission to Assess United States National Security Space Management and Organization* (Washington D.C., January 11, 2001) ix-x.

The President should consider establishing space as a national security priority.

2. Presidential Space Advisory Group

The President might find it useful to have access to high-level advice in developing a long-term strategy for sustaining the nation's role as the leading space-faring nation.

The President should consider the appointment of a Presidential Space Advisory Group to provide independent advice on developing and employing new space capabilities.

3. Senior Interagency Group for Space

The current interagency process is inadequate to address the number, range and complexity of today's space issues, which are expected to increase over time. A standing interagency coordination process is needed to focus on policy formulation and coordination of space activities pertinent to national security and to assure that representation in domestic and international fora effectively reflects U.S. national security and other space interests.

The President should direct that a Senior Interagency Group for Space be established and staffed within the National Security Council structure.

4. SecDef/DCI Relationship

The issues relating to space between the Department of Defense and the Intelligence Community are sufficiently numerous and complex that their successful resolution and implementation require a close, continuing and effective relationship between the Secretary of Defense and the Director of Central Intelligence.

The Secretary of Defense and the Director of Central Intelligence should meet regularly to address national security space policy, objectives and issues.

5. Under Secretary of Defense for Space, Intelligence and Information

Until space organizations have more fully evolved, the Office of the Secretary of Defense would benefit from having a senior-level official with sufficient standing to serve as the advocate for space within the Department. The Secretary of Defense would assign this official responsibility to oversee the Department's research and development, acquisition, launch and operation of its space, intelligence and information assets; coordinate the military intelligence activities within the Department; and work with the Intelligence Community on long-range intelligence requirements for national security.

An Under Secretary of Defense for Space, Intelligence and Information should be established.

6. Commander in Chief of U.S. Space Command and NORAD and Commander, Air Force Space Command

The Commander in Chief, U.S. Space Command should continue to concentrate on space as it relates to warfare in the mediums of air, land and sea, as well as space. His primary role is to conduct space operations and provide space-related services, to include computer network defense/attack missions in support of the operations of the other CINCs, and national missile defense. This broad and varied set of responsibilities as CINCSPACE will leave less time for his other assigned duties.

The Secretary of the Air Force should assign responsibility for the command of Air Force Space Command to a four-star officer other than CINCSPACE/CINCNORAD.

The Secretary of Defense should end the practice of assigning only Air Force flight-rated officers to the position of CINCSPACE and CINCNORAD to ensure that an officer from any Service with an understanding of combat and space could be assigned to this position.

7. Military Services

The Department of Defense requires space systems that can be employed in independent operations or in support of air, land and sea forces to deter and defend against hostile actions directed at the interests of the United States. In the mid term a Space Corps within the Air Force may be appropriate to meet this requirement; in the longer term it may be met by a military department for space. In the nearer term, a realigned, rechartered Air Force is best suited to organize, train and equip space forces.

The Air Force should realign headquarters and

*field commands to more effectively organize, train
and equip for prompt and sustained space operations.
Assign Air Force Space Command (AFSPC)
responsibility for providing the resources to execute
space research, development, acquisition and
operations, under the command of a four-star
general. The Army and Navy would still establish
requirements and develop and deploy space systems
unique to each Service.*

*Amend Title 10 U.S.C. to assign the Air Force
responsibility to organize, train and equip for prompt
and sustained offensive and defensive air and space
operations. In addition, the Secretary of Defense
should designate the Air Force as Executive Agent
for Space within the Department of Defense.*

8. Aligning Air Force and NRO Space Programs

*The Department of Defense and the Intelligence Community would benefit
from the appointment of a single official within the Air Force with authority
for the acquisition of space systems for the Air Force and the NRO based
on the "best practices" of each organization.*

*Assign the Under Secretary of the Air Force as the
Director of the National Reconnaissance Office.
Designate the Under Secretary as the Air Force
Acquisition Executive for Space.*

9. Innovative Research and Development

*The Intelligence Community has a need for revolutionary methods,
including but not limited to space systems, for collecting intelligence.*

*The Secretary of Defense and the Director of Central
Intelligence should direct the creation of a research,
development and demonstration organization to focus
on this requirement.*

*Competitive centers of innovation that actively pursue space-related
research, development and demonstration programs are desirable.*

*The Secretary of Defense should direct the Defense
Advanced Research Projects Agency and the
Services' laboratories to undertake development and
demonstration of innovative space technologies and*

138

systems for dedicated military missions.

10. Budgeting for Space

Better visibility into the level and distribution of fiscal and personnel resources would improve management and oversight of space programs.

The Secretary of Defense should establish a Major Force Program for Space.

The Commission believes that its recommendations, taken as a whole, will enable the U.S. to sustain its position as the world's leading space-faring nation. Presidential leadership and guidance, coupled with a more effective interagency process and especially with improved coordination between the Department of Defense and the Intelligence Community, are essential if the nation is to promote and protect its interests in space.[239]

---

[239] *Report of the Commission to Assess United States National Security Space Management and Organization* (Washington D.C., January 11, 2001) xxxi-xxxv.

Glossary

ACSC	Air Command and Staff College
ADCS	Active Defensive Counterspace
ADR	Annual Defense Review
ASAT	Anti-Satellite
AU	Air University
BMD	Ballistic Missile Defense
COTS	Commercial-off-the-shelf
DoD	Department of Defense
FSCL	Fire Support Coordination Line
GEO	Geostationary Earth Orbit
GPS	Global Positioning System
HEO	Highly Elliptical Orbit
ICBM	Inter-Continental Ballistic Missile
LEO	Low Earth Orbit
MEO	Medium Earth Orbit
NASA	National Aeronautics and Space Administration
NCA	National Command Authority
NMD	National Missile Defense
NTMV	National Technical Means of Verification
NORAD	North American Air Defense
NRO	National Reconnaissance Office
OCS	Offensive Counterspace
PDCS	Passive Defensive Counterspace
SAAS	School of Advanced Airpower Studies
SAC	Strategic Air Command

START	Strategic Arms Reduction Talks
US	United States
USAF	United States Air Force
USCINCSPACE	United States Commander-in-Chief Space
USSPACECOM	United States Space Command
USSR	Union of Soviet Socialist Republics

proposition. "1. A plan or scheme suggested for acceptance…5c. A statement containing only logical constants and having a fixed truth-value." —From The American Heritage Dictionary, Second College Edition (Boston, Houghton Mifflin Co, 1995) 994.

spacepower. The ability of a state or non-state actor to achieve its goals and objectives in the presence of other actors on the world stage through control and exploitation of the space environment.[240]

---

[240] James L. Hyatt, III et al., "Space Power 2010," Research Report 95-05 (Maxwell AFB, Ala.: Air Command and Staff College, May 1995), 5.

Bibliography

Books and Monographs

Allison, Graham T. Essence of Decision: Explaining the Cuban Missile Crisis. Boston: Little,   Brown & Co, 1971.

Arquilla, John and David Ronfeldt, eds. *In Athena's Camp: Preparing for Conflict in the Information Age.* Santa Monica, CA: RAND, 1997.

Bell, Thomas D. "Weaponization of Space: Understanding Strategic and Technological Inevitabilities," *Occasional Paper 6* Maxwell AFB, AL: Air War College Center for Strategy and Technology, January 1999.

Booth, Nicholas. *Space: The Next 100 Years.* New York: Orion Books, 1990.

Boushey Homer A. "Blueprints for Space," *Air University Quarterly Review* 11 (Spring 1959): 16-29.

Builder, Carl H. *The Icarus Syndrome: The Role of Air Power Theory in the Evolution and Fate of the U.S. Air Force.* New Brunswick, NJ: Transaction Publishers, 1994.

____. *The Masks of War: American Military Styles in Strategy and Analysis.* Baltimore: Johns Hopkins University Press, 1989.

Daso, Dik. Architects of American Air Supremacy: Gen. Hap Arnold and Dr. Theodore Von Kámán. Maxwell AFB, AL: Air University Press, 1997.

DeBlois, Bruce M. "Ascendant Realms: Characteristics of Airpower and Space Power," in Philip S. Meilinger, ed., *The Paths of Heaven: The Evolution of Airpower Theory.* Maxwell AFB, AL: Air University Press, 1997.

____., ed. *Beyond the Paths of Heaven: The Emergence of Space Power Thought.* Maxwell AFB, AL: Air University Press, September 1999.

Dehqanzada, Yahya A. and Ann M. Florini. *Secrets for Sale: How Commercial Satellite Imagery Will Change the World.* Washington, D.C.: Carnegie Endowment for International Peace, 2000.

Futrell, Robert Frank. *Ideas, Concepts, Doctrine: A History of Basic Thinking in the United   States Air Force, 1907-1964.* Maxwell AFB, AL: Air University Press, 1971; reprint,   New York: Arno Press, 1980.

____. *Ideas, Concepts, Doctrine: Basic Thinking in the United States Air Force, 1961-1984,*   Vol. II, Maxwell AFB, AL: Air University Press, December 1989.

Gonzales, Daniel. *The Changing Role of the U.S. Military in Space.* Santa Monica, CA: RAND, 1999.

Gray, Colin S. *American Military Space Policy: Information Systems, Weapons Systems and Arms Control.* Cambridge, MA: Abt Books, 1982.

____. Modern Strategy. Oxford: Oxford Press, 1999.

Hall, R. Cargill and Jacob Neufeld. *The U.S. Air Force in Space: 1945 to the Twenty-first Century.* Washington, D.C.: Air Force History and Museums Program, 1998.

Hammond, Grant T. "Paths to Extinction," White Papers 4 Special Studies. Maxwell AFB, AL: Air University Press, 1996

Hays, Peter L. James M. Smith, Alan R. Van Tassel, and Guy M. Walsh, eds. *Spacepower for a New Millennium: Space and U.S. National Security.* New York: McGraw-Hill, 2000.

Hays, Peter L. "Struggling Towards Space Doctrine: U.S. Military Space Plans, Programs, and Perspectives during the Cold War," unpublished Ph.D. dissertation, Fletcher School of Law and Diplomacy, Tufts University, May 1994.

Jelonek, Mark P. "Toward and Air and Space Force: Naval Aviation and the Implications for Space Power," *Cadre Paper 5*. Maxwell AFB, AL: Air University Press, September 1999.

Johnson, Dana J., Scott Pace, and C. Bryan Gabbard. *Space: Emerging Options for National Power*. Santa Monica: RAND National Defense Research Institute, 1998.

Keaney, Thomas A. and Eliot A. Cohen. *Revolution in Warfare? Air Power in the Persian Gulf.* Annapolis: Naval Institute Press, 1995.

Klotz, Frank G. *Space, Commerce, and National Security*. New York: Council on Foreign Relations Press, 1998.

Lupton, David E. *On Space Warfare: A Space Power Doctrine.* Maxwell AFB, Ala.: Air University Press, 1998.

Luttwak, Edward N. *Strategy: The Logic of War and Peace.* Cambridge: Harvard University Press, Belknap Press, 1987.

Mantz, Michael R. *The New Sword: A Theory of Space Combat Power.* Maxwell AFB, AL: Air University Press, May 1995.

McDougall, Walter A. . . . *the Heavens and the Earth:* A Political History of the Space Age. New York: Basic Books, 1985.

Meilinger, Col Philip. *Ten Propositions Regarding Airpower.* Washington, D.C.: Air Force History and Museums Program, 1995.

Meilinger, Col Phillip. *The Paths of Heaven.* School of Advanced Air Power Studies, Maxwell AFB, Ala.: Air University Press, 1997.

Muolo, Michael, ed. *Space Handbook: A Warfighter's Guide to Space* Vol. 1 Maxwell AFB, AL: Air University Press, 1993.

Neufeld, Jacob, George M. Watson, Jr., and David Chenoweth, eds. *Technology and the Air Force: A Retrospective Assessment.* Washington, D.C.: Air Force History and Museums Program, 1997.

Neufeld, Michael J. *The Rocket and the Reich: Peenemünde and the Coming of the Ballistic Missile Era.* New York: Free Press, 1995.

Oberg, James E. *Space Power Theory.* Washington, D.C.: GPO, March 1999.

Paret, Peter, ed. *The Makers of Modern Strategy: From Machiavelli to the Nuclear Age.* Princeton: Princeton University Press, 1986.

Peebles, Curtis. *Battle for Space.* New York: Beaufort Books, 1983.

_____. *High Frontier: The U.S. Air Force and the Military Space Program.* Washington, D.C.: Air Force History and Museums Program, 1997.

Possel, William H. "Lasers and Missile Defense: New Concepts for Space-Based and Ground-Based Laser Weapons," Occasional Paper 5. Maxwell AFB, AL: Air War College Center for Strategy and Technology, July 1998.

Report of the Commission to Assess National Security Space Management and Organization. Washington, D.C.: Commission to Assess National Security Space Management and Organization, 11 January 2001.

Report of the Independent Commission on the National Imagery and Mapping Agency. *The Information Edge: Imagery Intelligence and Geospatial Information in an*

Evolving National Security Environment. Washington, D.C.: Independent Commission on the National Imagery and Mapping Agency, December 2000.

Report of the National Commission for the Review of the National Reconnaissance Office. *The NRO at the Crossroads.* Washington, D.C.: National Commission for the Review of the National Reconnaissance Office, 1 November 2000.

Spacy, William L., II. "Does the United States Need Space-Based Weapons?" *Cadre Paper 4.* Maxwell AFB, AL: Air University Press, September 1999.

Spires, David N. *Beyond Horizons: A Half Century of Air Force Space Leadership.* Peterson AFB, CO: Air Force Space Command, 1997.

Weigley, Russell F. *The American Way of War: A History of United States Military Strategy and Policy.* Bloomington: Indiana University Press, 1973.

Waltz, Kenneth N. *Man, the State, and War: A Theoretical Analysis.* New York: Columbia University Press, 1954.

Watts, Barry D. *The Military Use of Space: A Diagnostic Assessment.* Washington, D.C.: Center for Strategic and Budgetary Assessments, February 2001.

Wolfe, Tom. *The Right Stuff.* New York: Bantam Press, 1980.

U.S. Military Uses of Space, 1945-1991: Index and Guide. Washington, D.C.: The National Security Archive and Alexandria, VA: Chadwyck-Healey, Inc., 1991.

Book Chapters and Articles

Anson, Peter. "First Space War: The Contribution of Satellites to the Gulf War." *RUSI Journal*, Winter 1991, 45-53.

Baum, Michael E. "Defiling the Altar: The Weaponization of Space," *Airpower Journal* 8 (Spring 1994): 52-61.

Behling, Thomas G. and Kenneth McGruther. "Satellite Reconnaissance of the Future," *Joint Forces Quarterly* 18 (Spring 1998): 23-30.

Budura, Victor P. "The Next Force," *Air Chronicles* (June 1998).

Burrows, William E. "Satellite Reconnaissance," in James E. Dillard and Walter T. Hitchcock, eds. *The Intelligence Revolution and Modern Warfare.* Chicago: Imprint Publications, 1996, 183-199.

Cohen, William S. "Department of Defense Directive 3100.10: Space Policy," Washington, D.C.: Office of the Secretary of Defense, 9 July 1999.

Covault, Craig. "Military Space Dominates Air Strikes," *Aviation Week & Space Technology* (29 March 1999): 31-33.

_____. "Recon, GPS Operations Critical to NATO Strikes," *Aviation Week & Space Technology* (26 April 1999): 35-37.

_____. "USAF Shifts Technology for New Future in Space," *Aviation Week & Space Technology* (17 August 1998): 40-44.

Daso, Dik. "New World Vistas: Looking toward the Future, Learning from the Past," *Aerospace Power Journal* 13 (Winter 1999): 67-76.

Day, Dwayne A. "Listening from Above: The First Signals Intelligence Satellite," *Spaceflight* 41 (August 1999): 338-346.

Dublois, Bruce M. "Space Sanctuary: A Viable National Strategy," *Airpower Journal* 12 (Winter 1998): 41-57.

Fulghum, David A. "Space Beckons Future AWACS," *Aviation Week & Space Technology* (21 September 1998): 62-63.

Hays, Peter L. and Karl P. Mueller. "Going Boldly—Where? Aerospace Integration, the Space Commission, and the Air Force's Vision for Space," *Aerospace Power Journal* 15 no. 1 (Spring 2001): 34-49.

Hays, Peter L. and Roy F. Houchin, II. "Commercial Spysats and Shutter Control: The Military Implications of U.S. Policy on Selling and Restricting Commercial Remote Sensing Data," unpublished paper prepared for the USAF Institute for National Security Studies, 1 October 1999.

Hoffman, Tim. "Space Plays Key Role in Operation Desert Fox," Air Force Space Command Public Affairs release.

Houchin, Roy F., II. "The Rise and Fall of Dyna-Soar: The USAF's First Hypersonic Program," *The Proceedings of the National Aerospace Conference.* Dayton, OH: Wright State University Press, forthcoming.

Gray, Colin S. and John B. Sheldon. "Spacepower and the Revolution in Military Affairs: A Glass Half Full?" *Airpower Journal* 13 (Fall 1999): 23-38.

Grier, Peter. "Partners in Space," *Air Force Magazine* 82 (February 1999): 28-32.

Jennings, Frank W. "Aerospace and Air and Space," *Airpower Journal* 10 (Spring 1996): 117-118.

Lambeth, Benjamin S. "Air Power, Space Power, and Geography," *Journal of Strategic Studies* 22 (June/September 1999): 63-82.

_____. "The Synergy of Air and Space," *Airpower Journal* 12 (Summer 1998): 4-14.

Moorman, Thomas S., Jr. "The Explosion of Commercial Space and the Implications for National Security," *Airpower Journal* 13 (April 1999): 6-20.

National Science and Technology Council. "Fact Sheet: National Space Policy," Washington, D.C.: The White House, 19 September 1996.

Owens, William. "The Emerging U.S. Systems of Systems," in Stuart E. Johnson and Martin C. Libicki, eds. *Dominant Battlespace Knowledge: The Winning Edge.* Washington, D.C.: NDU Press, 1995.

Power, John. "Space Control in the Post-Cold War Era," *Airpower Journal* 4 (Winter 1990): 24-33.

Rife, Maj Shawn P. "On Space Power Separatism." Air Command and Staff College Student Paper: Maxwell AFB, Ala, April 1998.

Scott, William B. "Airbreathing HyperSoar Would 'Bounce' on Upper Atmosphere," *Aviation Week & Space Technology* (7 September 1998): 126-130.

_____. "US Adopts 'Tactical' Space Control Policy," *Aviation Week & Space Technology* (29 March 1999): 35.

_____. "Wargames Zero In on Knotty Milspace Issues," *Aviation Week & Space Technology* (29 January 2001): 40-42.

Smith, Bob. "The Challenge of Space Power, *Airpower Journal* 13 (April 1999): 32-40.

SPACECAST 2020 White Paper, "Space Lift: Suborbital, Earth to Orbit, and on Orbit," *Airpower Journal*, n.p. on-line, Internet, available from http://www.airpower.maxwell.af.mil/airchronicles/apj/spacast3.html.

Tobin, David M. "Man's Place in Space-Plane Flight Operations: Cockpit, Cargo Bay, or Control Room?" *Airpower Journal* 13 (Fall 1999): 50-65.

United States Department of State. "Treaty on Principles Governing the Activities of States in the Exploration and Use of Outer Space, Including the Moon and Other Celestial Bodies (Outer Space Treaty)," Washington, D.C.: Department of State, 27 January 1967.

_____. "Treaty Between the United States of America and the Union of Soviet Socialist Republics on the Limitation of Anti-Ballistic Missile Systems (ABM Treaty)," Washington, D.C.: Department of State, 26 May 1972.

Worden, Simon Peter. "The Air Force and Future Space Directions: Are We Good Stewards?" *Aerospace Power Journal*. 15, no. 1 (Spring 2001): 50-55.

Government Documents

A National Security Strategy for a New Century. The White House, 1998.

Air Force Doctrine Document (AFDD) 1: Air Force Basic Doctrine. HQ AFDC/DR, September, 1997.

AFDD 2: Organization and Employment of Aerospace Power. HQ AFDC/DR, 17 February, 2000.

AFDD 2-2: Space Operations. HQ AFDC/DR, 23 August 1998.

AFDD 2-5: Information Operations. HQ AFDC/DR, 5 August, 1998.

AFDD 4. "Space Operations Doctrine." 10 July 1996.

College of Aerospace Doctrine, Research, and Education (CADRE). 2025. Maxwell AFB, Ala.: Air University Press, August, 1996.

Department of the Air Force. Global Presence 1995. Washington, D.C.: Government Printing Office (GPO), 1995.

SPACECAST 2020 Executive Summary. Maxwell AFB, Ala.: Air University Press, June, 1994.

USSPACECOM. Operation Desert Shield and Desert Storm Assessment. Peterson AFB, Colo.: USSPACECOM, 31 January 1992.

Briefings/Speeches/Lectures

Donahue, Lt Gen William J. "Air Force Communications" Briefing to the students of Air Command and Staff College, Maxwell AFB, Ala, February 2000.

Estes III, Gen Howell M., USCINCSPACE. "Convergence of Space Sectors—A New Symbiosis." Speech given to the 1998 National Space Symposium, US Space Foundation, 9 April, 1998.

_____., USCINCSPACE. "Doctrinal Lineage of Space." Speech given to the Air Force Association in Los Angeles, 18 October, 1996.

_____., USCINCSPACE. "The Air Force at a Crossroad." Speech given to the Air Force Association, 14 November, 1997.

Hall, Keith R., Director of the National Reconnaissance Office. Keynote address to the Office of Special Projects Holiday Party, 1 December, 1997.

_____., Director of the National Reconnaissance Office. "Space Policy, Programs, and Operations." Presentation to the Committee on Armed Services Subcommittee on Strategic Forces United States Senate. Washington, D.C., 22 March 1999.

Meyers, Gen Richard B., USCINCSPACE. "Achieving the Promise of Space—The Next Step." Speech given to the Air Force Association Warfighting Symposium, 4 February 1999.

Short, Lt Gen Michael C. "ALLIED FORCE—Airpower's Victory." Briefing to the students of Air Command and Staff College, Maxwell AFB, Ala, August, 1999.

Internet Sources

Grier, Peter. "Partners in Space." *Air Force Magazine*, February 1999. On-line. Internet. Available from http://www.afa.org/magazine/0299partners.html.

Johnson, Bryan T. "The New Space Race: Challenges for U.S. National Security and Free Enterprise." The Heritage Foundation Backgrounder, 23 August 1999. On-line. Internet. Available from http://heritage.org/library/backgrounder/bg1316.html.

Katzman, SMSgt Jim. "Space Systems Support Joint Forces Near Iraq." Air Force News Service, 6 March 1998. On-line. Internet. Available from http://www.af.mil/news/Mar1998/n19980306_980308.html.

Leonard, David. "US Air Force Eyes Fully Integrated Aerospace Force." Spacenews.com, 20 January 2000. On-line. Internet. Available fromhttp://www.spacenews.com/stemp/crossroads/crossroads_3.html.

NATO Press Conference by Mr Jaimie Shea, NATO Spokesman, 28 May 1999 (during ALLIED FORCE). On-line. Internet. Available from http://www.eucom.mil/operations/af/nato/1999/may/99may28.htm.

Newman, Richard J. "The New Space Race." U.S. News On-line, 8 November 1999. On-line. Internet. Available from http://www.usnews.com/usnews/issue/991108/space.htm.

Plaxco, Jim. "Owen Garriott Reflections of Space Exploration." *Planetary Studies Foundation*, 1998. On-line. Internet. Available from http://www.planets.org/ogi.htm.

"Control of Space Key to Future War." Spacedaily, 10 May 1999. On-line. Internet. Available from http://www.spacedaily.com/spacecast/news/milspace-99c.html.

"Space Role in ALLIED FORCE Extensive—Effective." USSPACECOM News Release 11-99. HQ USSPCAECOM/PA, 17 June 1999. On-line. Internet. Available from http://www.spacecom.af.mil/usspace/new11-99.htm.

www.ingramcontent.com/pod-product-compliance
Lightning Source LLC
Chambersburg PA
CBHW050619110426
42813CB00010B/2614